Collins

AF735204

Maths Progress Tests for White Rose

Year 2/P3

Cherri Moseley

William Collins' dream of knowledge for all began with the publication of his first book in 1819. A self-educated mill worker, he not only enriched millions of lives, but also founded a flourishing publishing house. Today, staying true to this spirit, Collins books are packed with inspiration,
innovation and practical expertise. They place you at the centre of a world of possibility and give you exactly what you need to explore it.

Collins. Freedom to teach.

Collins
An imprint of HarperCollins*Publishers*
The News Building
1 London Bridge Street
London
SE1 9GF

Browse the complete Collins catalogue at
www.collins.co.uk

© HarperCollinsPublishers Limited 2019

10 9 8 7 6 5 4 3 2 1

ISBN 978-0-00-833351-5

All rights reserved. No part of this publication may be reproduced, stored in a retrieval system,
or transmitted in any form by any means, electronic, mechanical, photocopying, recording or otherwise, without the prior written permission of the Publisher or a licence permitting restricted copying in the United Kingdom issued by the Copyright Licensing Agency Ltd., Barnard's Inn, 86 Fetter Lane, London, EC4A 1EN.

British Library Cataloguing in Publication Data. A catalogue record for this publication is available from the British Library.

Author: Cherri Moseley
Publisher: Katie Sergeant
Commissioning Editor: Fiona Lazenby
Product Developer: Mike Appleton
Copyeditor: Joan Miller
Proofreader: Catherine Dakin
Design and Typesetting: Ken Vail Graphic Design
Cover Design: The Big Mountain Design
Production controller: Katharine Willard
Printed and bound by CPI Group (UK) Ltd, Croydon, CR0 4YY

FSC www.fsc.org
MIX
Paper from responsible sources
FSC™ C007454

This book is produced from independently certified FSC™ paper to ensure responsible forest management.

For more information visit: **www.harpercollins.co.uk/green**

Contents

How to use this book

Introduction

Collins Maths Progress Tests for White Rose have been designed to give you a consistent whole-school approach to teaching and assessing mathematics. Each photocopiable book covers the required mathematics objectives from the 2014 Primary English National Curriculum. For teachers in Scotland, the books can offer guidance and structure that is not provided in the Curriculum for Excellence Experiences and Outcomes or Benchmarks for Numeracy and Mathematics.

As stand-alone tests, the *Collins Maths Progress Tests for White Rose* provide a structured way to assess progress in arithmetic and reasoning skills, to help you identify areas for development, and to provide evidence towards expectations for each year group. Whilst the tests are independent of any textbook-based teaching and learning scheme to allow for maximum flexibility, the content for each test has been selected based on the suggested teaching order in the *White Rose Maths Schemes of Learning*, which are designed to support a mastery approach to teaching and learning.

Assessment of mathematical skills

At the end of KS1 and KS2, children sit tests to assess the standards they have reached in mathematics. This is done through national curriculum tests (SATs) in Arithmetic and Mathematical Reasoning. *Collins Maths Progress Tests for White Rose* have been designed to provide children with opportunities to explore a range of question types whilst building familiarity with the format, language and style of the SATs.

The Arithmetic tests comprise constructed response questions, presented as context-free calculations, to assess pupils' confidence with a range of mathematics operations as appropriate to the year group. Questions come from the Number, Ratio and Algebra domains.

The Reasoning tests present mathematical problems in a wide range of formats to ensure pupils can fully demonstrate mathematical fluency, mathematical problem solving and mathematical reasoning. They include both selected response questions (e.g. multiple choice, matching, yes/no) and constructed response questions. Questions may draw on all content domains and approximately half of the questions in the Reasoning tests are presented in context.

The tests follow the structure and format of SATs mathematics papers and are pitched at a level appropriate to age-related expectations for the year group. They provide increasing challenge within each year group and across the school, both in terms of content and cognitive demand, but also with increasing numbers of questions to build stamina and resilience. Using the progress tests with your classes at the end of each half-term should help pupils to develop and practise the necessary skills required to complete the national tests with confidence, as well as offering you a snapshot of their progress at those points throughout the year. You can use the results formatively to help identify gaps in knowledge and next teaching steps.

How to use this book

In this book, you will find twelve photocopiable tests: one arithmetic test and one reasoning test for use at the end of each half term of teaching. Each child will need a copy of the test. You will find Curriculum Content Coverage on page vi indicating the White Rose Scheme of Learning Block and associated Content Domain topics covered in each test across the year group. The specific Content Domain references indicating the year, strand and substrand, e.g. 2N1, for the questions in each test are in the tables on page 139. You may find it useful to make a photocopy of these tables for each child and highlight questions answered incorrectly to help identify any consistent areas of difficulty.

The number of marks available and suggested timing to be allowed are indicated for each test. The number of marks/questions in each test and the length of time allowed increases gradually across the year as summarised in the table below. Note that the Year 2 and Year 6 Summer term tests have been written as full practice papers assuming that all content will have been taught by this point. They mirror the number of marks and time allowed in the end of Key Stage 1 and end of Key Stage 2 test papers.

Year group	Test	Time allowed	Number of marks
2	Autumn 1 Arithmetic	15 minutes	15
2	Autumn 1 Reasoning	25 minutes	20
2	Autumn 2 Arithmetic	15 minutes	15
2	Autumn 2 Reasoning	28 minutes	25
2	Spring 1 Arithmetic	18 minutes	20
2	Spring 1 Reasoning	25 minutes	25
2	Spring 2 Arithmetic	18 minutes	20
2	Spring 2 Reasoning	30 minutes	30
2	Summer 1 Arithmetic	20 minutes	25
2	Summer 1 Reasoning	35 minutes	35
2	Summer 2 Arithmetic	20 minutes	25
2	Summer 2 Reasoning	35 minutes	35

To help you mark the tests, you will find mark schemes at the back of the book. These include the answer requirement, number of marks to be awarded, additional guidance on answers that should or should not be accepted and when to award marks for working in multi-mark questions.

Test demand

The tests have been written to assess progress in children's arithmetic and mathematical reasoning skills with the content and cognitive demand of questions increasing within each book and across the series to build towards to end of key stage expectations of the SATs. Since the national tests may cover content from the whole key stage, each progress test contains some questions which draw on content from earlier terms or previous year objectives (particularly in autumn term tests). This ensures that prior content and skills are revisited.

The level of demand for each question has been provided within the mark schemes for each test using the notation T (working towards), E (expected standard) or G (greater depth). These ratings are given as an indication of the level of complexity of each question taking into account the thinking skills required to understand what is being asked, the computational complexity in calculating the answer, spatial reasoning or data interpretation required and the response strategy for the question.

Performance thresholds

The table below provides guidance for assessing how children perform in the tests. Most children should achieve scores at or above the expected standard, with some children working at greater depth and exceeding expectations for their year group. While the thresholds bands do not represent standardised scores, as in the end of key stage SATs, they will give an indication of how pupils are performing against the expected standards for their year group. The thresholds have been set broadly assuming that pupils who achieve greater than 60% will be working at the expected standard and those who score more than 80% are likely to be working at greater depth. However, pupils will all have individual strengths and weaknesses, so it is possible that they could be working towards the expected standard in some areas but at greater depth in others. For this reason, using the content domain coverage tables to identify common areas of difficulty alongside your own professional judgement, will enable you to identify pupils' specific gaps in knowledge and areas where further teaching may be required.

Tracking progress

A record sheet is provided to help you illustrate to children the areas in which their arithmetic and reasoning skills are strong and where they need to develop. A spreadsheet tracker is also provided via collins.co.uk/assessment/downloads which enables you to identify whole-class patterns of attainment. This can be used to inform your next teaching and learning steps.

Editable download

All the files are available online in Word and PDF format. Go to collins.co.uk/assessment/downloads to find instructions on how to download. The files are password protected and the password clue is included on the website. You will need to use the clue to locate the password in your book.

You can use these editable files to help you meet the specific needs of your class, whether that be by increasing or decreasing the challenge, by reducing the number of questions, by providing more space for answers or increasing the size of text for specific children.

Year group	Test	Working towards (T)	Expected standard (E)	Greater depth (G)
2	Autumn 1 Arithmetic	8 marks or below	9–11 marks	12–15 marks
2	Autumn 1 Reasoning	11 marks or below	12–15 marks	16–20 marks
2	Autumn 2 Arithmetic	8 marks or below	9–11 marks	12–15 marks
2	Autumn 2 Reasoning	14 marks or below	15–19 marks	20–25 marks
2	Spring 1 Arithmetic	11 marks or below	12–15 marks	16–20 marks
2	Spring 1 Reasoning	14 marks or below	15–19 marks	20–25 marks
2	Spring 2 Arithmetic	11 marks or below	12–15 marks	16–20 marks
2	Spring 2 Reasoning	17 marks or below	18–23 marks	24–30 marks
2	Summer 1 Arithmetic	14 marks or below	15–19 marks	20–25 marks
2	Summer 1 Reasoning	20 marks or below	21–27 marks	28–35 marks
2	Summer 2 Arithmetic	14 marks or below	15–19 marks	20–25 marks
2	Summer 2 Reasoning	20 marks or below	21–27 marks	28–35 marks

Curriculum content coverage

All content objectives from the Year 2 National Curriculum Programme of Study for Mathematics are covered within one or more of the half-termly progress tests across the year. The content for each test is based on the suggested teaching order of the White Rose Maths Schemes of Learning. The table below shows from which teaching blocks the content for each test is drawn. Where the White Rose Maths blocks are devoted to skills or consolidation rather than introduction of new content, these blocks are not covered by the tests. The Summer tests for Year 2 draw on content from previous blocks.

White Rose Schemes of Learning blocks			Collins Maths Progress Tests for White Rose											
Blocks	**Weeks**	**Topics**	**Autumn 1: Arithmetic**	**Autumn 1: Reasoning**	**Autumn 2: Arithmetic**	**Autumn 2: Reasoning**	**Spring 1: Arithmetic**	**Spring 1: Reasoning**	**Spring 2: Arithmetic**	**Spring 2: Reasoning**	**Summer 1: Arithmetic**	**Summer 1: Reasoning**	**Summer 2: Arithmetic**	**Summer 2: Reasoning**
Autumn Block 1	Weeks 1–3	Number: Place Value	✔	✔							✔	✔	✔	✔
Autumn Block 2	Weeks 4–6	Number: Addition and Subtraction	✔	✔							✔	✔	✔	✔
	Weeks 7–8				✔	✔								
Autumn Block 3	Weeks 9–10	Measurement: Money				✔								
Autumn Block 4	Weeks 11–12	Number: Multiplication and Division			✔	✔					✔	✔	✔	✔
Spring Block 1	Weeks 1–2	Number: Multiplication and Division					✔	✔			✔	✔	✔	✔
Spring Block 2	Weeks 3–4	Statistics						✔						
Spring Block 3	Weeks 5–6	Geometry: Properties of Shape						✔						
	Week 7									✔				
Spring Block 4	Weeks 8–10	Number: Fractions							✔	✔	✔	✔	✔	✔
Spring Block 5	Week 11	Measurement: Length and height								✔				
Spring Block 6	Week 12	Consolidation												
Summer Block 1	Weeks 1–3	Geometry: Position and direction										✔		
Summer Block 2	Weeks 4–5	Problem Solving and efficient methods												
Summer Block 3	Week 6	Measurement: Time										✔		
	Week 7													✔
Summer Block 4	Weeks 8–10	Measurement: Mass, Capacity and Temperature												✔
Summer Block 5	Weeks 11–12	Investigations												

Year 2 Autumn Half Term 1: Arithmetic Name ____________

1	70 – 30 =
	1 mark
2	100 – 70 = ☐
	1 mark

© HarperCollinsPublishers Ltd 2019

3 $\frac{1}{2}$ of 8 = ☐

1 mark

4 89 + 1 = ☐

1 mark

© HarperCollinsPublishers Ltd 2019

5 $17 - 7 =$ ☐

1 mark

6 $2 + 40 + 2 =$ ☐

1 mark

© HarperCollins*Publishers* Ltd 2019

7 50 ÷ 10 = ☐

1 mark

8 5 + 43 = ☐

1 mark

© HarperCollins*Publishers* Ltd 2019

9 37 + ☐ = 44

1 mark

10 86 – ☐ = 36

1 mark

© HarperCollins*Publishers* Ltd 2019

 Name ____________

11 96 + 6 = ☐

1 mark

12 64 – 7 = ☐

1 mark

© HarperCollins*Publishers* Ltd 2019

13 2 + ☐ + 3 = 13

1 mark

14 ☐ = 18 – 3

1 mark

© HarperCollinsPublishers Ltd 2019

15 100 = ☐ + 40

1 mark

Total marks ………/15

© HarperCollinsPublishers Ltd 2019

 Name ____________

1 How many cookies are there altogether?

[ ] cookies

1 mark

2 Which number is represented below?

[ ]

1 mark

3 Tick the odd one out.

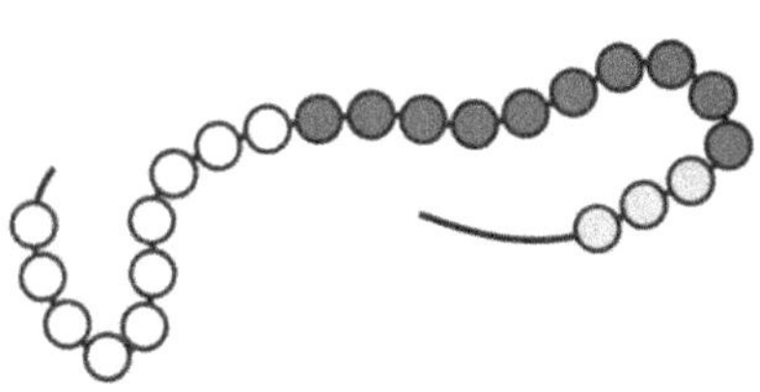

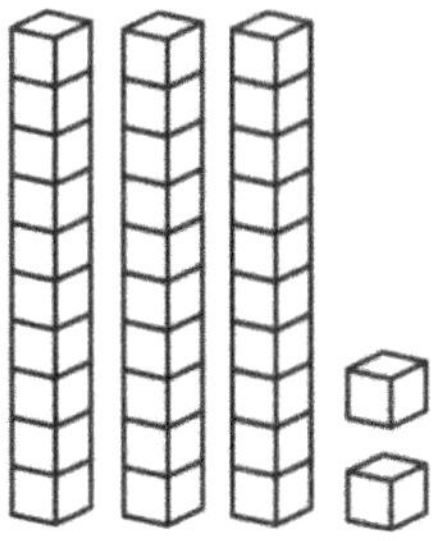

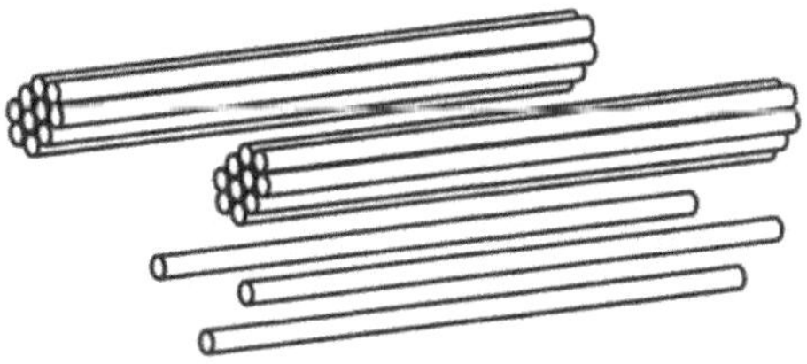

1 mark

© HarperCollins*Publishers* Ltd 2019

4 Here is part of a number line.

Write the correct number in each box.

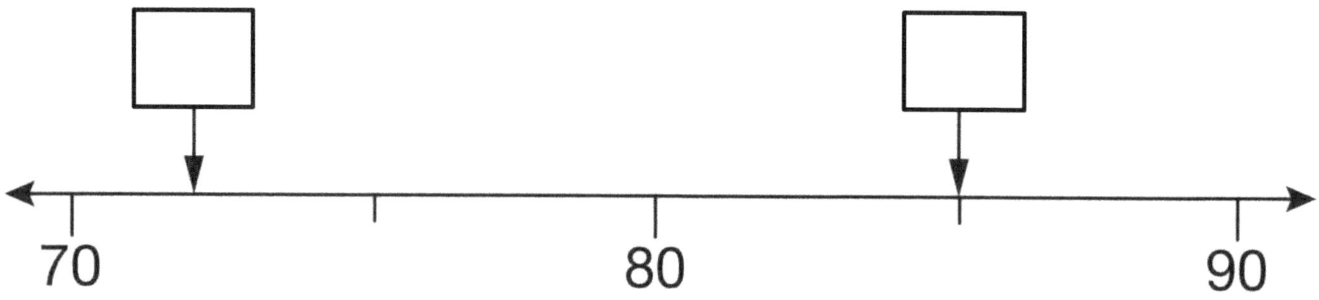

2 marks

5 Match the numerals to the words.

60	16	61	19

sixty-one	nineteen	sixty	sixteen

1 mark

6 Amy makes a pattern with sandcastles.

Some are plain and some have turrets.

She writes a number pattern on the flags.

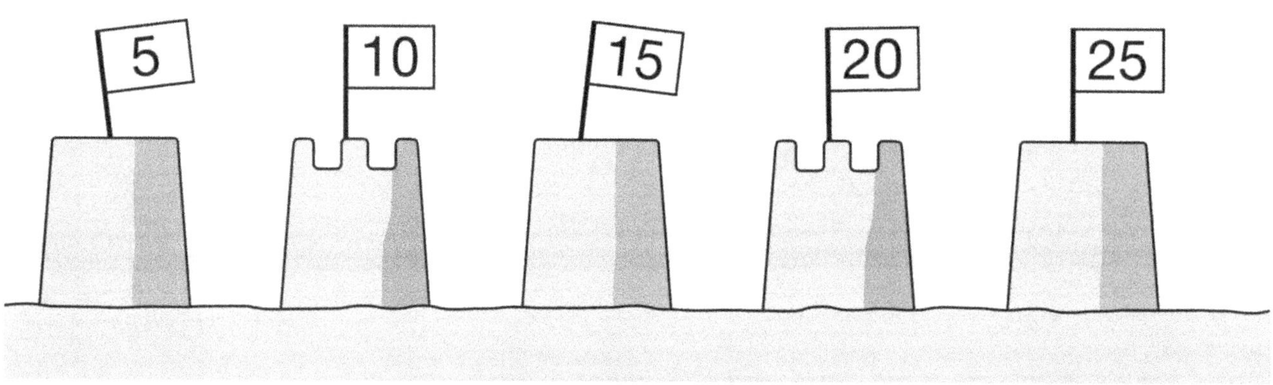

Write the number that will be on the flag of the next **plain** castle.

1 mark

© HarperCollins*Publishers* Ltd 2019

Name ______________

7

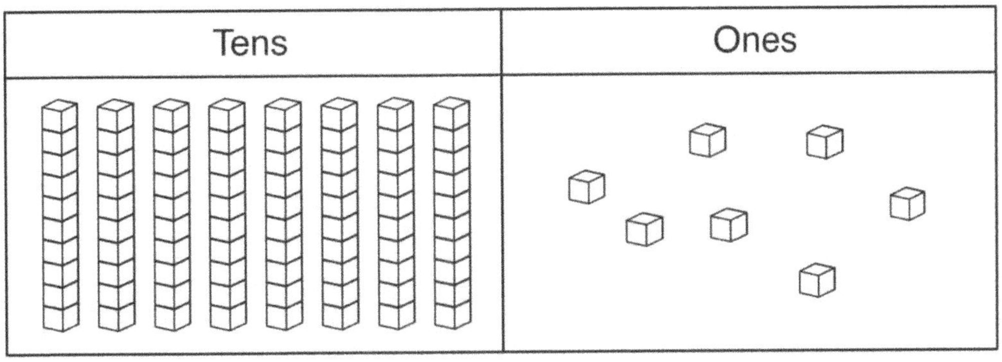

What number is shown in the place value chart?

☐

1 mark

8 Write these numbers in order, starting with the smallest.

92 29 42 94 49

☐ ☐ ☐ ☐ ☐

smallest largest

1 mark

9 Match the sums which have the same answer.

One is done for you.

30 + 6	30 + 26
50 + 6	70 + 26
70 + 6	10 + 26
90 + 6	50 + 26

1 mark

© HarperCollins*Publishers* Ltd 2019

 Name ____________

10 The numbers on this number line go up by the **same amount** each time.

Write the missing numbers in the boxes.

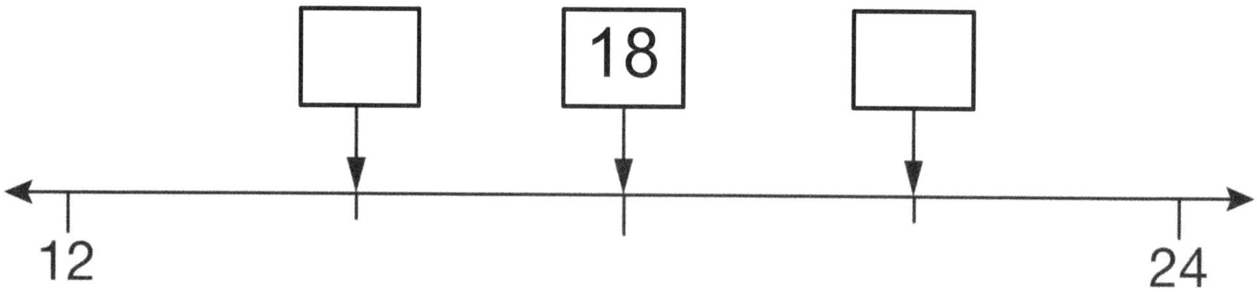

1 mark

11 Emma has 42 marbles.

She loses 7 marbles in a game.

How many marbles does she have left?

marbles

1 mark

12 Ali partitions some numbers. Complete his diagram below.

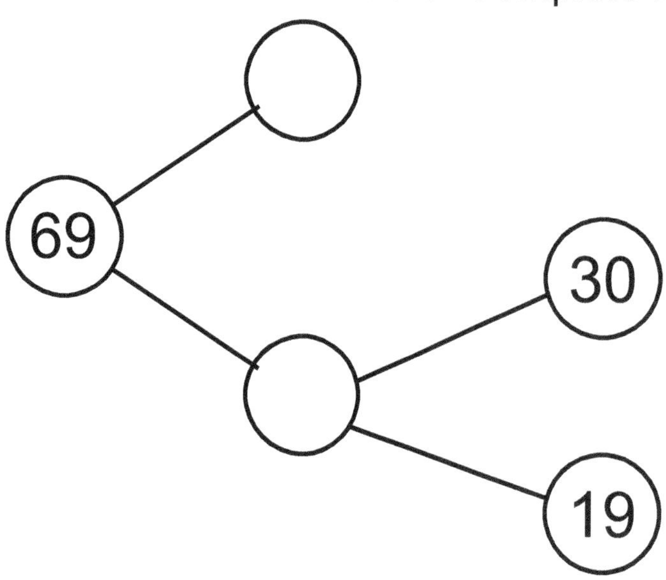

1 mark

© HarperCollins*Publishers* Ltd 2019

13 Look at these numbers.

24 37 52 71

Write each number **once** to make these correct.

☐ < ☐

☐ > ☐

1 mark

14 Sam makes 24 using different shapes for tens and ones.

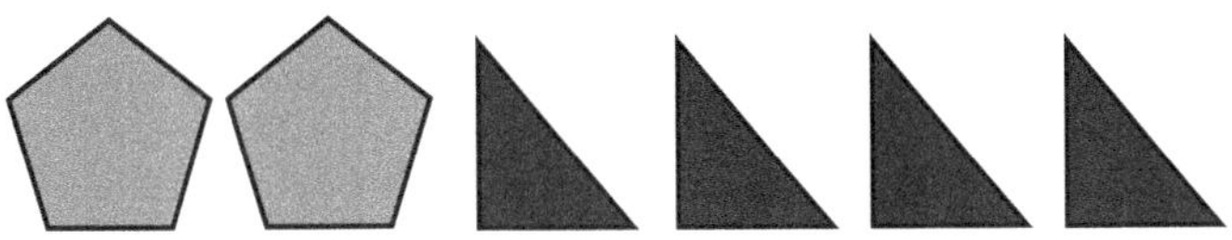

Sam makes a new number.

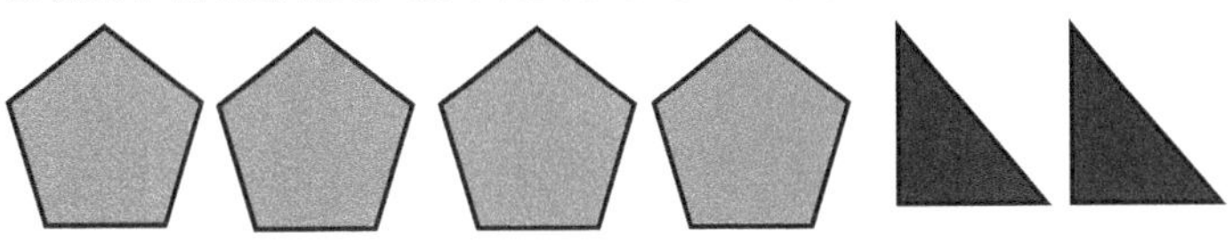

What is Sam's new number?

☐

1 mark

15 There are **100** pencils in a box.

There are **30** children in Cedar Class and **20** children in Ash class.

Each child is given a pencil.

How many pencils are **left**?

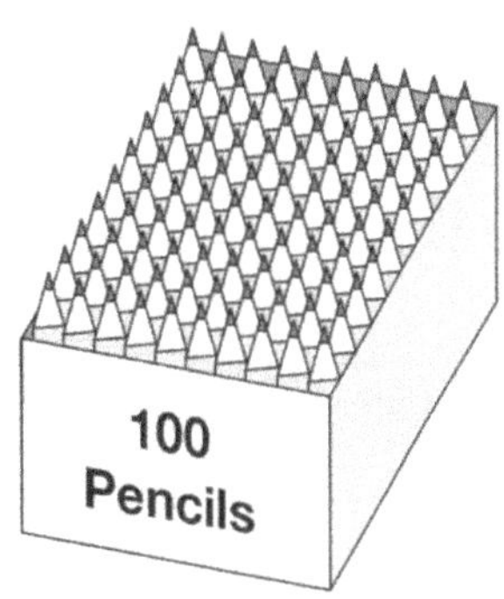

Show your method

pencils

2 marks

© HarperCollins*Publishers* Ltd 2019

16 Complete these calculations.

One is done for you.

10 – 9 = 1

100 – ☐ = 10

100 – ☐ = 90

1 mark

17

<table>
<tr><td colspan="2">51</td></tr>
<tr><td>10</td><td>41</td></tr>
</table>

Use the same three numbers, 10, 41 and 51 to make two different number sentences.

One is done for you.

41 + 10 = 51

☐ – ☐ = ☐

☐ = ☐ – ☐

1 mark

18 Write the missing number in the box.

43 + 10 = ☐ – 10

1 mark

Total marks ………/20

© HarperCollins*Publishers* Ltd 2019

1 $20 + 30 + 40 =$ ☐

1 mark

2 $39 + 40 =$ ☐

1 mark

© HarperCollins*Publishers* Ltd 2019

3 $11 - 5 =$ ☐

1 mark

4 $7 \times 5 =$ ☐

1 mark

© HarperCollins*Publishers* Ltd 2019

 Name ____________

5	78 – 10 – 10 = ☐
	1 mark
6	62 + ☐ = 69
	1 mark

© HarperCollins*Publishers* Ltd 2019

7 47 + 28 = ☐

1 mark

8 44 – 7 = ☐

1 mark

© HarperCollinsPublishers Ltd 2019

9 9 × ☐ = 90

1 mark

10 11 × 2 = ☐

1 mark

© HarperCollinsPublishers Ltd 2019

11 $7 + 8 + 9 = 7 + \square + 8$

1 mark

12 $100 - 57 = \square$

1 mark

© HarperCollins*Publishers* Ltd 2019

13 [ ] = 19 – 7

1 mark

14 9 × 3 = [ ]

1 mark

© HarperCollinsPublishers Ltd 2019

15 $4 \times 10 = \square \times 5$

1 mark

Total marks ………/15

© HarperCollinsPublishers Ltd 2019

 Name ____________

1 Circle the **two** numbers that are odd.

57 62

91 34

1 mark

2 How many crayons altogether?

[ ] crayons

1 mark

3 Amir and Sarah added **two consecutive** one-digit numbers.

They each made a **total** of **17**.
Which **two** numbers did they add?

[ ]

1 mark

© HarperCollins*Publishers* Ltd 2019

4 Complete these sums.

3 + 7 = ☐ 13 + 7 = ☐ 30 + 70 = ☐

1 mark

5 Shakira has these coins in her purse.
Complete the part–whole diagrams.

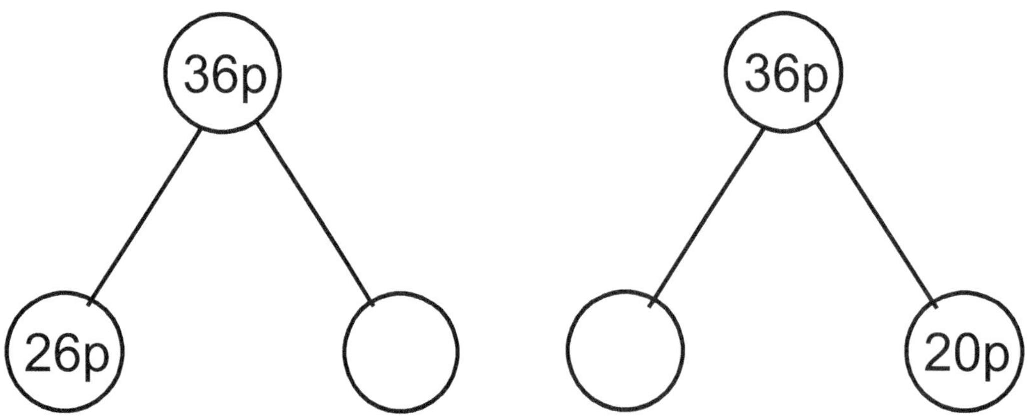

1 mark

6 At the cake shop, **5** cupcakes are put in each box.
How many cupcakes are in the boxes **altogether**?

☐ cupcakes

1 mark

© HarperCollins*Publishers* Ltd 2019

7 Lucy has **52p**.
James has 20p **less** than Lucy.
Tick the coins **James** has.

1 mark

8 Look at the array.

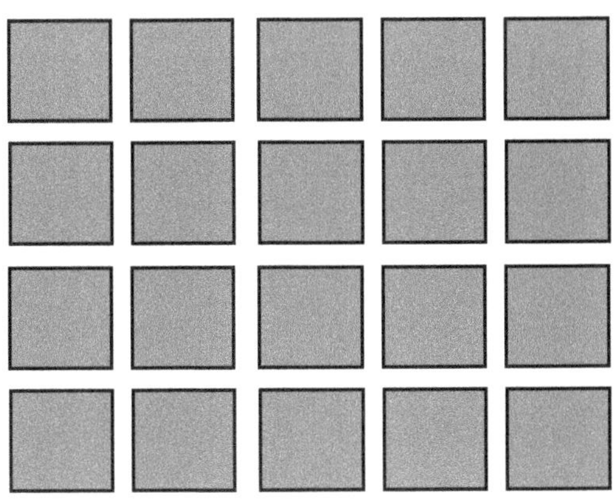

Now look at the four calculations below.

Tick (✓) the calculation that describes the array.

5 + 4	5 × 4
5 – 4	5 + 5 + 5

1 mark

9 Write the **same** number in both boxes to make the sum correct.

☐ + ☐ = 18

1 mark

10 Write six **different** numbers to make these sums correct.

☐ + ☐ = 51

☐ + ☐ = 51

☐ + ☐ = 51

2 marks

11 One length of a swimming pool is **10** metres.
Ben swims the length of the pool **6** times.
Ben works out how many metres he swims altogether.
Circle the **two** calculations that Ben can use.

10 + 6

6 × 10

6 + 6 + 6 + 6 + 6 + 6

10 + 10 + 10 + 10 + 10 + 10

1 mark

© HarperCollinsPublishers Ltd 2019

12 Fill in the missing numbers to make each pair of cards **total 19**.
One pair is done for you.

10	9
	7
	14

1 mark

13 Write <, = or > in each circle to make the statements correct.

10 × 5 ◯ 12 × 2

10 × 2 ◯ 5 × 5

1 mark

14 Erin has **67** stickers.
She has **19 more** than Tom.
How many stickers does **Tom** have?

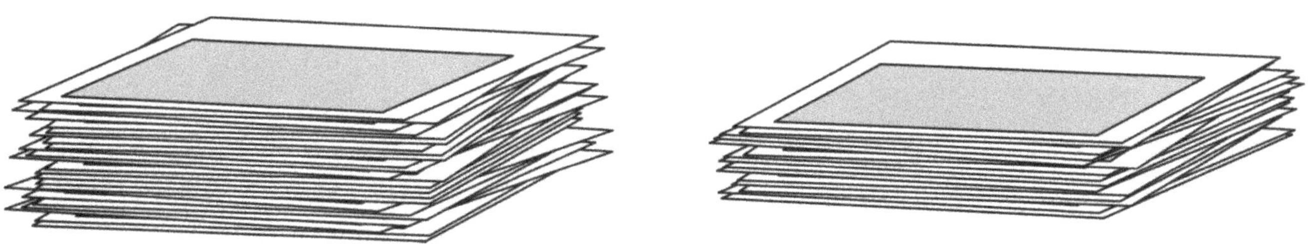

[ ] stickers

1 mark

© HarperCollins*Publishers* Ltd 2019

15 Mark each statement with a tick (✓) or a cross (✗).

One has been done for you.

54 < 84 – 30 ✗

67 > 70 + 7

85 = 90 – 5

1 mark

16 This is all the money Sita has saved to buy a pair of trainers.

How much more money does Sita need to save?

£

1 mark

17 Dad blew up **42** balloons for the party.
17 balloons burst.
How many balloons are **left** for the party?

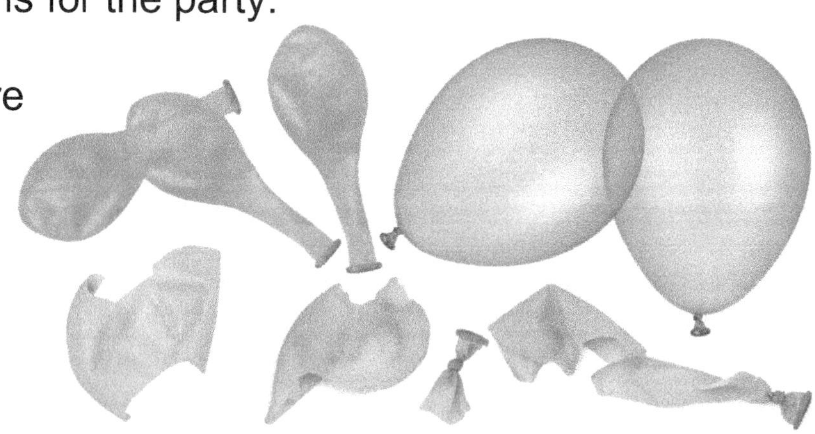

balloons

1 mark

© HarperCollins*Publishers* Ltd 2019

18 Ben has **50p** in 5p coins.
How many 5p coins does Ben have?

[ ] coins

1 mark

19 Emma buys a banana for **15p**.
She pays with a **50p** coin.
How much **change** does Emma get?

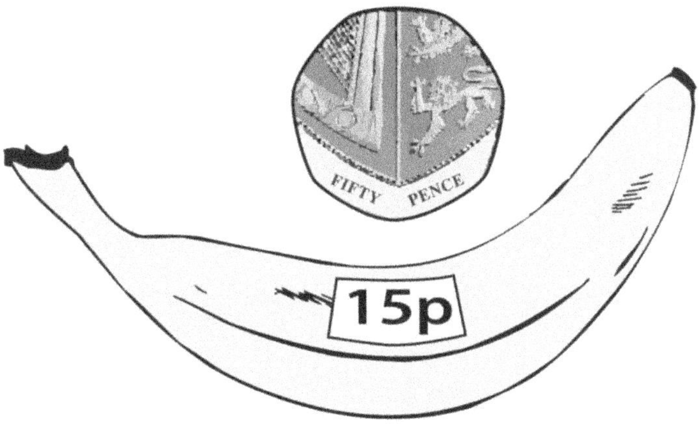

[ ] p

1 mark

20 Here is an array of **30** squares.

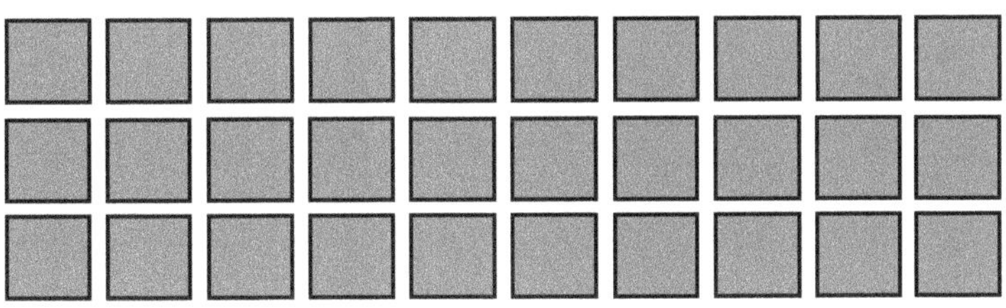

Write two multiplication sentences to match the array.

[ ] × [ ] = 30

[ ] × [ ] = 30

1 mark

21

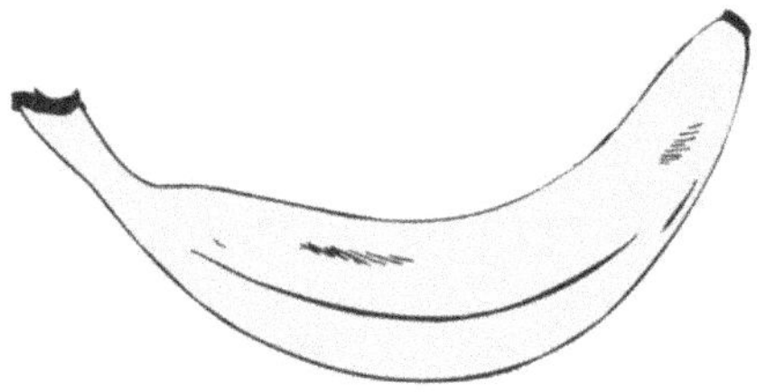

bananas 15p each	apples 20p each

Sam buys **2** bananas and **3** apples.
How much does Sam spend **altogether**?

Show your method

p

2 marks

22 Write each addition sentence as a multiplication sentence.

3 + 2 + 5 + 5 + 5 + 5 + 5 + 5 = 35

☐ × ☐ = 35

10 + 5 + 10 + 5 + 10 = 40

☐ × ☐ = 40

1 mark

23 Write the missing number to make this number sentence correct.

27 + 7 – ☐ = 19

1 mark

Total marks ………/25

© HarperCollins*Publishers* Ltd 2019

1 40 + 30 = ☐

1 mark

2 3 + 50 + 3 = ☐

1 mark

© HarperCollinsPublishers Ltd 2019

3 $9 \times 10 =$ ☐

1 mark

4 $18 \div 2 =$ ☐

1 mark

© HarperCollins*Publishers* Ltd 2019

5 7 + 8 + 9 = ☐

1 mark

6 3 × 5 = ☐

1 mark

© HarperCollinsPublishers Ltd 2019

7 $6 + 7 =$ ☐

1 mark

8 $55 \div 5 =$ ☐

1 mark

© HarperCollins*Publishers* Ltd 2019

9 64 – 19 = ☐

1 mark

10 7 + ☐ + 6 = 16

1 mark

© HarperCollinsPublishers Ltd 2019

 Name ____________

11 7 × 2 = ☐

1 mark

12 52 + 48 = ☐

1 mark

© HarperCollinsPublishers Ltd 2019

13 60 + [ ] = 90

1 mark

14 70 ÷ [ ] = 7

1 mark

© HarperCollinsPublishers Ltd 2019

15 52 – 17 = ☐

1 mark

16 ☐ = 19 – 5

1 mark

© HarperCollins*Publishers* Ltd 2019

17 97 + 10 = ☐

1 mark

18 37 + 23 = ☐ + 37

1 mark

© HarperCollinsPublishers Ltd 2019

 Name ____________

19 $\square - 16 = 31$

1 mark

20 $10 \times 2 = 4 \times \square$

1 mark

Total marks ………/20

© HarperCollins*Publishers* Ltd 2019

 Name ____________

1 Ben puts **10** biscuits on each plate.
He uses **8** plates.

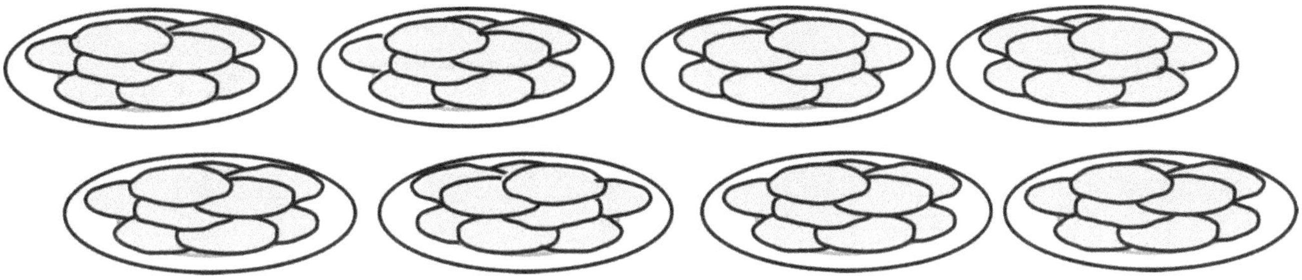

Complete the multiplication calculation to show how many biscuits are on the plates **altogether**.

☐ × ☐ = ☐

1 mark

2

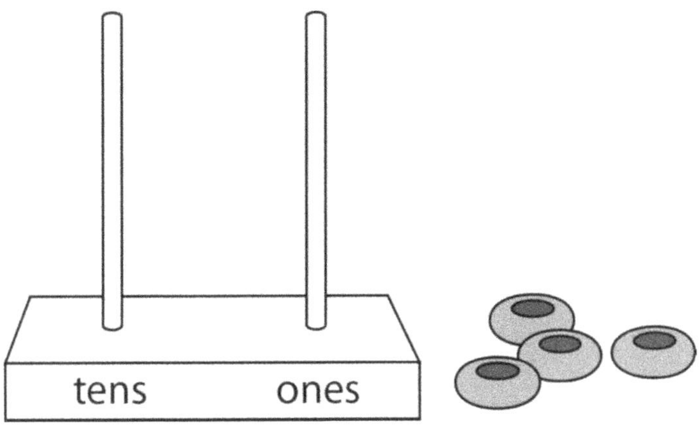

There are **4** beads for the abacus.
Which **5** numbers can be made if you use all the beads?
Write the numbers in the boxes in any order.

☐ ☐ ☐ ☐ ☐

1 mark

© HarperCollins*Publishers* Ltd 2019

3 Tick (✓) the names of the **two** shapes in this picture.

Tick (✓) **two**.

- triangle ☐
- square ☐
- rectangle ☐
- circle ☐
- hexagon ☐

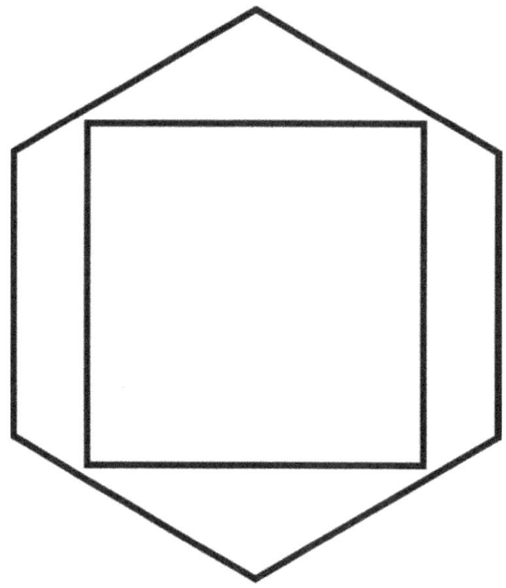

1 mark

4 Share these marbles equally among **5** children.

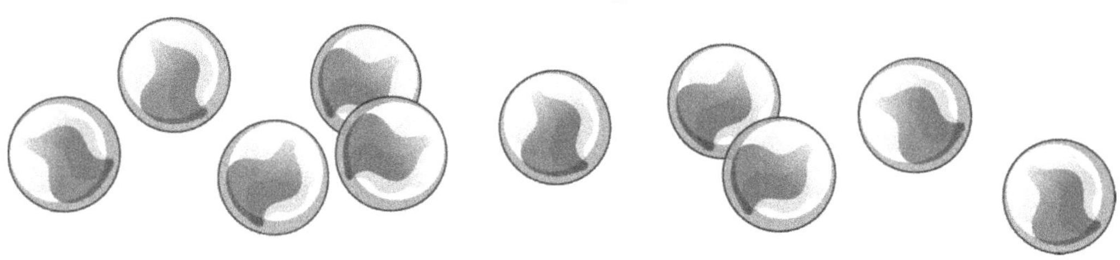

How many marbles does each child get?

[ ] marbles

1 mark

5 Match the sums that have the same answer.
One is done for you.

17 + 8	30 + 15
37 + 8	70 + 15
57 + 8	10 + 15
77 + 8	50 + 15

1 mark

© HarperCollins*Publishers* Ltd 2019

Name ____________

6 Write the **same** number in both boxes to make the sum correct.

$\square + \square = 48$

1 mark

7 Ben buys a toy car for 90p.

Tick (✓) three coins to show how to make 90p.

Tick (✓) four coins to show how to make 90p.

2 marks

8 Use only these numbers to make a different number sentence each time.

5   9   45

One has been done for you.

$9 \times 5 = 45$

$\square \div \square = \square$

$\square \times \square = \square$

1 mark

© HarperCollinsPublishers Ltd 2019

 Name ____________

9 Complete the number sentences to show **three different** ways to partition the set of counters.

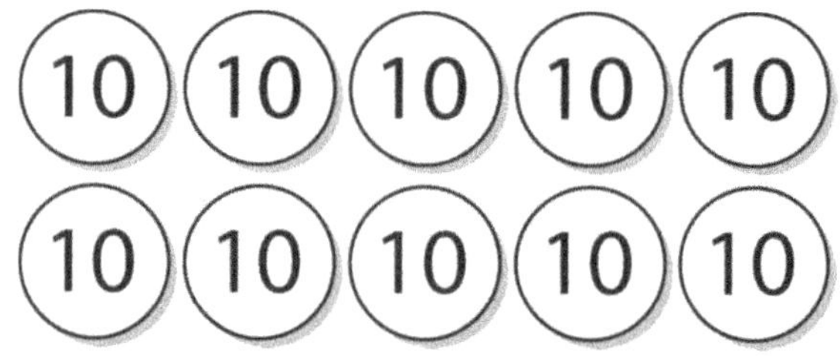

□ + □ + □ = □

□ + □ + □ = □

□ + □ + □ = □

1 mark

10 A classroom has **12** tables.

Each table has **2** children sitting at it.

Complete the number sentence to show how many children there are **altogether**.

□ × □ = □ children

1 mark

11 Erin's mum is **37** years old.
Erin's grandmother is **63** years old.
How much **older** is Erin's grandmother than Erin's mum?

□ years

1 mark

© HarperCollins*Publishers* Ltd 2019

 Name ____________

12 Tomas has **3** coins in his pocket.
Each coin is worth less than 20p.
Tomas does **not** have **3** coins that are the **same**.
What is the most money Tom could have?

[p]

1 mark

13 Children in Elm class chose their favourite fruit.
The block diagram shows the results.

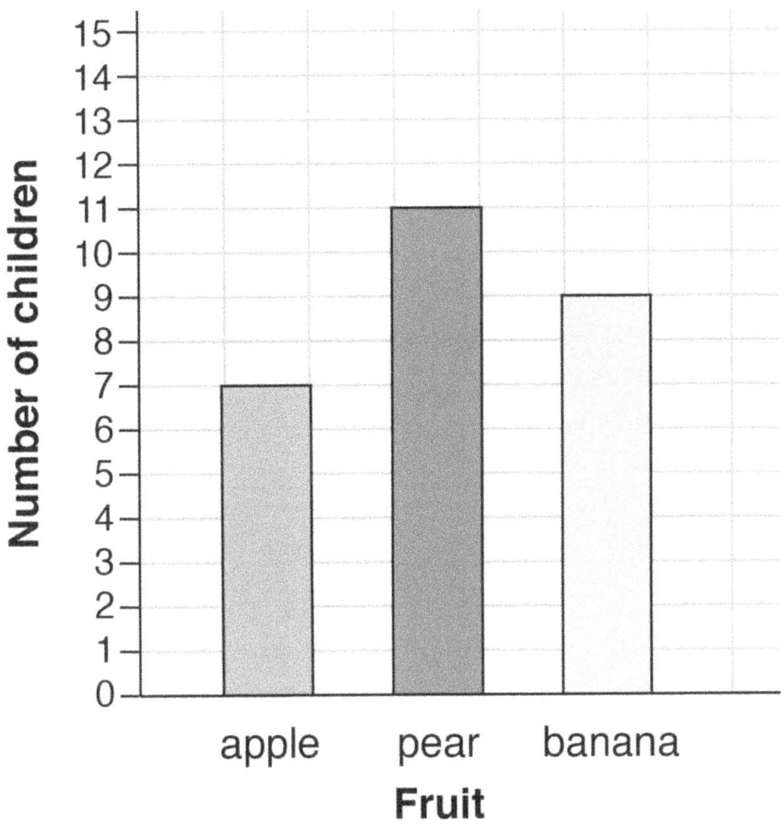

a) How many **more** children chose pear than chose apple?

[children]

1 mark

b) How many children **altogether** chose their favourite fruit?

[children]

1 mark

© HarperCollins*Publishers* Ltd 2019

14 Tick (✓) the **hexagon**.

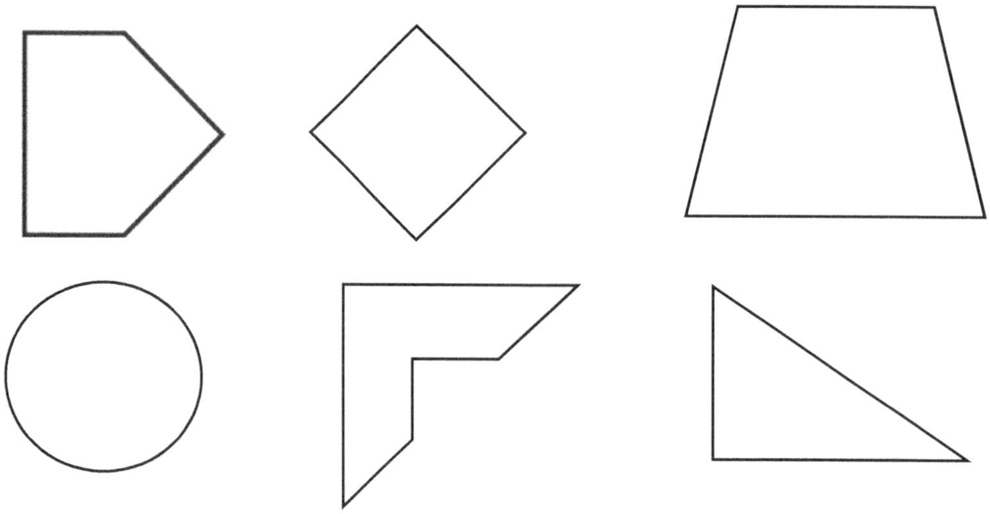

1 mark

15 What is the **price** of the trainers?

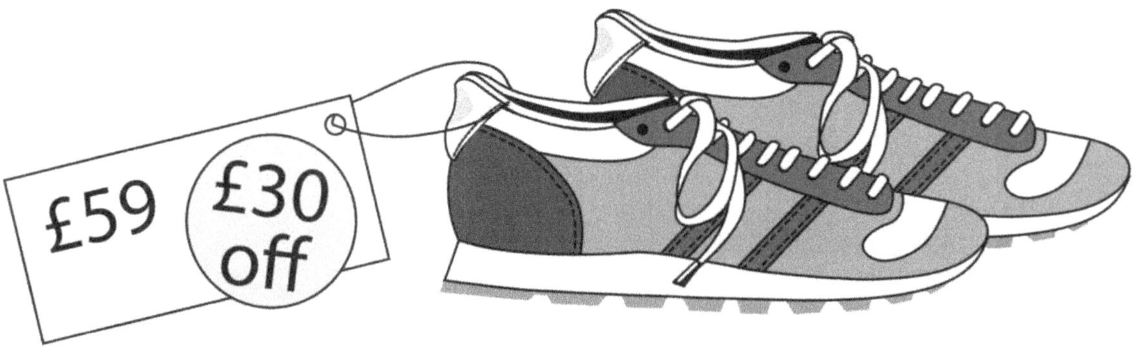

Show your method

£

1 mark

16 Write the correct sign <, = or > in each box.

74 □ 56 + 18

87 □ 56 + 17

62 □ 56 + 16

1 mark

17 Asma has a packet of **36** pens.
She takes **9** pens out of the packet.
How many pens are **left** in the packet?

□ pens

1 mark

18 An orange costs **18p**.
Ben buys **1** orange.
Ben pays with a **50p** coin.
Tick (✓) the change Ben gets.

1 mark

19 Which of these 3D shapes has **at least one** triangular face?
Tick (✓) the names of the possible shapes.

cube	square-based pyramid
cylinder	cuboid
triangular prism	sphere

1 mark

© HarperCollins*Publishers* Ltd 2019

 Name ____________

20 Use **four different** numbers to complete these calculations.

☐ × ☐ = 30 ☐ × ☐ = 30

1 mark

21 There are **57** cars in the car park.
16 more cars go into the car park.
Then **45** cars leave the car park.
How many cars are in the car park **now**?

Show your method

cars

2 marks

22

34	
16	18

Which **two** inverse calculations could you use to check 16 + 18 = 34?

Tick (✓) the two inverse calculations.

34 – 18 = 16 18 + 16 = 34

18 – 16 = 34 34 – 16 = 18

1 mark

Total marks ………/25

© HarperCollins*Publishers* Ltd 2019

 Name ____________

1 40 + 20 + 10 = ☐

1 mark

2 80 – 50 = ☐

1 mark

© HarperCollinsPublishers Ltd 2019

3 $8 \times 2 = \square$

1 mark

4 $1 + \square + 5 = 15$

1 mark

© HarperCollinsPublishers Ltd 2019

5 $8 + 11 = \square$

1 mark

6 $\frac{1}{2}$ of $18 = \square$

1 mark

© HarperCollinsPublishers Ltd 2019

7 71 – 18 = ☐

1 mark

8 38 + 25 = ☐

1 mark

© HarperCollinsPublishers Ltd 2019

9 6 × 10 = ☐

1 mark

10 27 + 73 = ☐

1 mark

© HarperCollins*Publishers* Ltd 2019

11 $\frac{1}{3}$ of 15 = ☐

1 mark

12 91 – 28 = ☐

1 mark

© HarperCollins*Publishers* Ltd 2019

13 $7 \times 3 =$ ☐

1 mark

14 $35 \div 5 =$ ☐

1 mark

© HarperCollins*Publishers* Ltd 2019

15 54 + ☐ = 94

1 mark

16 49 + ☐ = 28 + 49

1 mark

© HarperCollins*Publishers* Ltd 2019

17 30 ÷ ☐ = 10

1 mark

18 ☐ – 22 = 36

1 mark

© HarperCollinsPublishers Ltd 2019

 Name ____________

19 $3 \times 10 = 6 \times$ ☐

1 mark

20 $\frac{3}{4}$ of 12 = ☐

1 mark

Total marks ………/20

© HarperCollins*Publishers* Ltd 2019

1 Circle the **two** numbers that are even.

17 54

82 39

1 mark

2 Put the four snakes in order, from longest to shortest.

One is done for you.

☐ ☐ A ☐

longest shortest

1 mark

3

Tick the **fraction** of the strip that is shaded.

$\frac{1}{2}$ ☐ $\frac{1}{4}$ ☐

$\frac{1}{3}$ ☐ $\frac{3}{4}$ ☐

1 mark

© HarperCollins*Publishers* Ltd 2019

4 Complete the table.

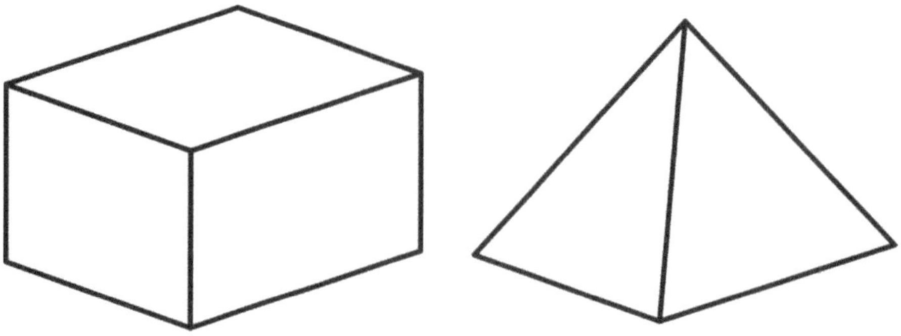

3D shapes	**faces**	**vertices**	**edges**
cuboid		8	
square-based pyramid			8

1 mark

5 Write the **same** number in both boxes to make the sum correct.

☐ + ☐ = 52

1 mark

6 Tick the box that does not show $\frac{1}{4}$.

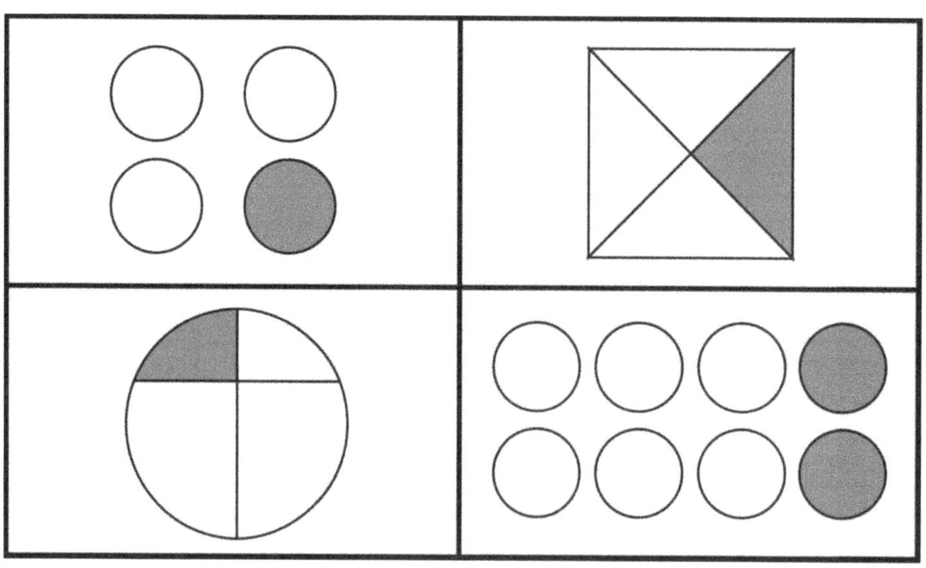

1 mark

© HarperCollins*Publishers* Ltd 2019

7 Ben's sunflower was 27 cm tall.
It grows 8 cm taller.
How tall is Ben's sunflower now?

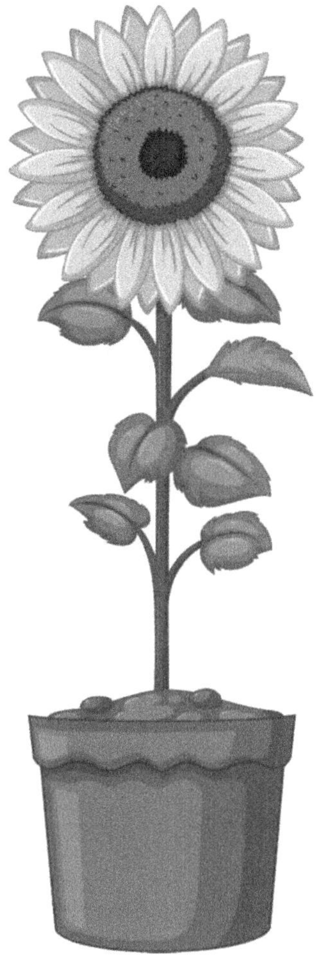

1 mark

8 The tally chart shows how many stickers each child has collected.

Sam	𝍸 𝍸 III
Ana	𝍸 𝍸 𝍸 II
Erin	𝍸 𝍸
Ali	𝍸 𝍸 IIII

a) Sam has fewer stickers than Ana.
How **many** fewer?

1 mark

b) Which children have collected **fewer** stickers than Ali?

1 mark

© HarperCollins*Publishers* Ltd 2019

9 Shade $\frac{3}{4}$ of this shape.

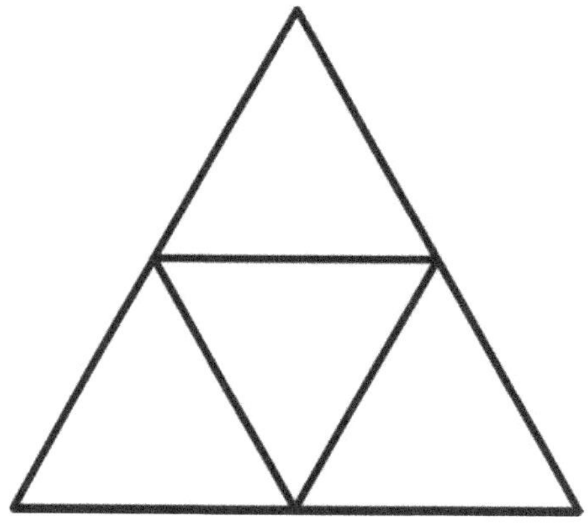

1 mark

10 Write six **different** numbers to make these sums correct.

$\square + \square = 72$

$\square + \square = 72$

$\square + \square = 72$

2 marks

11 Sanya has **already** saved this amount of money to buy a bicycle.

How much **more** money does Sanya need to save?

£ ______

1 mark

© HarperCollins*Publishers* Ltd 2019

 Name ____________

12 Jo and Raj counted bricks.
Jo counted **34** blue bricks.
Raj counted **27** yellow bricks.
How many bricks did they count **altogether**?

[ ] bricks

1 mark

13 A class has **70** coloured pencils.
There are **10** pots to put them in.
Each pot must have the same number of coloured pencils.
How many coloured pencils should go in each pot?

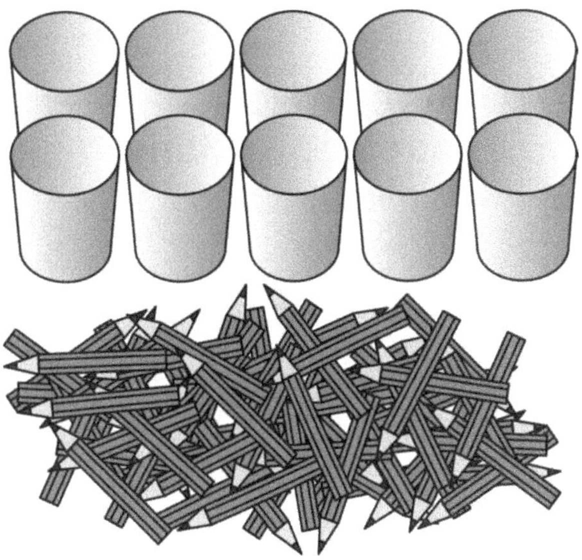

[ ] coloured pencils

1 mark

14 Tick the shape that does **not** have a line of symmetry.

1 mark

© HarperCollinsPublishers Ltd 2019

15 Match each object to its likely length.
One has been done for you.

length of a bed	30 centimetres
length of a ruler	5 metres
length of a car	2 centimetres
length of a paperclip	2 metres

1 mark

16 Ali is **6** years old.
His dad's age is **5 times** Ali's age.
How old is Ali's dad?

[ ] years old

1 mark

17 Here is part of a fraction number line.

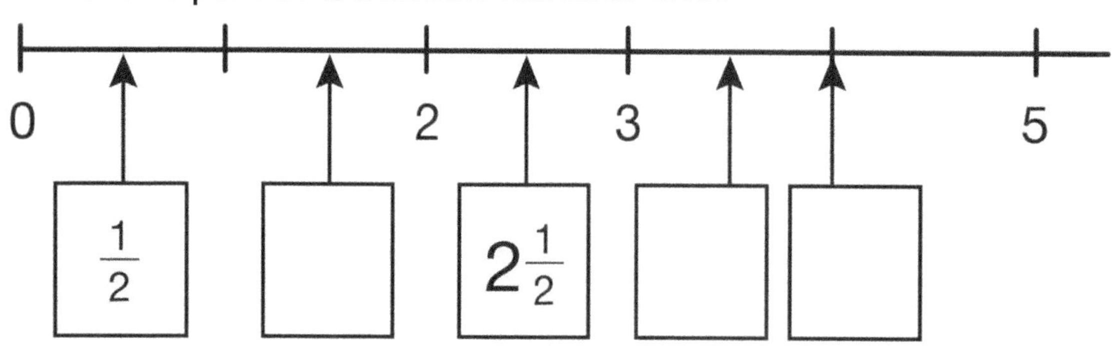

Write the correct number in each box.

1 mark

© HarperCollins*Publishers* Ltd 2019

 Name ______________

18 Alex has £1 in 10p coins.
How **many** 10p coins does Alex have?

coins

1 mark

19 Draw a **square** with sides **6 cm** long.
Use a ruler.

1 mark

© HarperCollins*Publishers* Ltd 2019

 Name ____________

20 Shakira had some toy figures.
She gave **half** of them to Erin.
Then she gave **half of what she had left** to Laura.
She has **3** toy figures left.

How many toy figures did Shakira start with?

Show your method

______ toy figures

2 marks

21 Use **only** these numbers to make a different number sentence each time.

18	2	9

One is done for you.

9	×	2	=	18
	×		=	
	÷		=	

1 mark

© HarperCollinsPublishers Ltd 2019

22 Lisa has **75p**.
Emma has **40p** less than Lisa.
Tick the coins that Emma has.

1 mark

23 A piece of string is **98 cm** long.

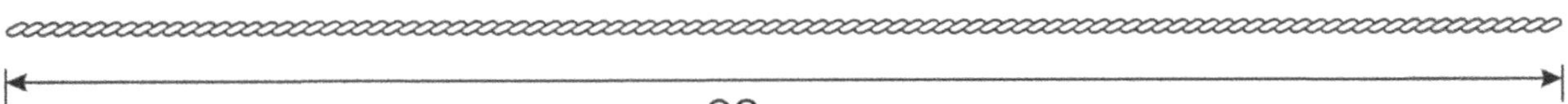

Ben cut off **45 cm**.
How much string is left?

[ ] centimetres

1 mark

24 Write the correct sign <, = or > in each box.

3 metres [ ] 30 centimetres

105 cm [ ] 1 m 15 cm

100 cm [ ] 1 metre

1 mark

© HarperCollins*Publishers* Ltd 2019

25 Tom has **24 more** football cards than Ben.
Ben has **36** football cards.
How many football cards do Tom and Ben have **altogether**?

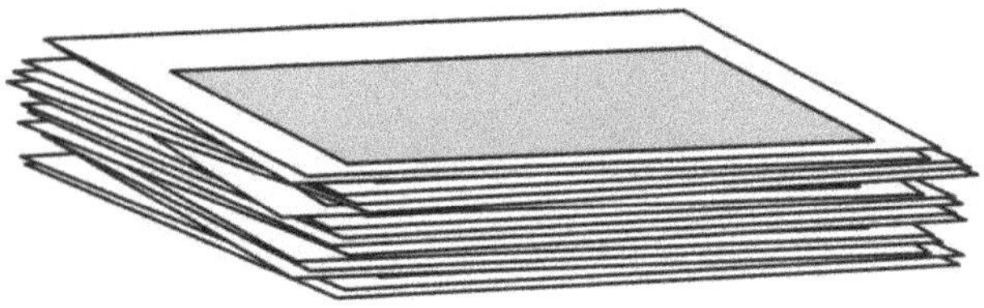

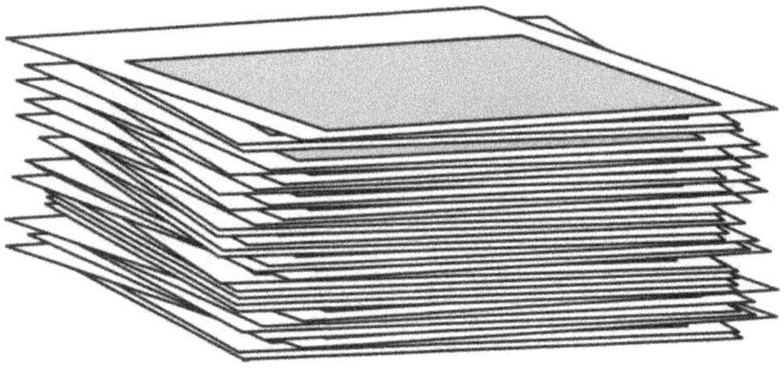

[ ] football cards

1 mark

26 Write the missing number to make this number sentence correct.

24 + 18 – [ ] = 12

1 mark

27 Use the **inverse** to write a checking number sentence for this calculation.

45 – 29 = 16

[ ]

1 mark

Total marks ………/30

© HarperCollins*Publishers* Ltd 2019

1 5 + 3 = ☐

1 mark

2 10 + 20 + 30 + 40 = ☐

1 mark

© HarperCollins*Publishers* Ltd 2019

3 51 – 8 ☐

1 mark

4 6 + 7 + 8 = ☐

1 mark

© HarperCollins*Publishers* Ltd 2019

5 $11 \times 5 =$ ☐

1 mark

6 $100 - 68 =$ ☐

1 mark

© HarperCollins*Publishers* Ltd 2019

7 20 – ☐ = 13

1 mark

8 20 + ☐ = 90

1 mark

© HarperCollins*Publishers* Ltd 2019

9	$\frac{1}{2}$ of 26 = ☐
	1 mark
10	70 – 0 = ☐
	1 mark

© HarperCollins*Publishers* Ltd 2019

 Name ____________

11 74 + 8 = ☐

1 mark

12 35 ÷ 5 = ☐

1 mark

© HarperCollinsPublishers Ltd 2019

13 48 + 52 = ☐

1 mark

14 92 – 37 = ☐

1 mark

© HarperCollins*Publishers* Ltd 2019

15 22 ÷ 2 = ☐

1 mark

16 96 + 7 = ☐

1 mark

© HarperCollinsPublishers Ltd 2019

17 $\frac{1}{3}$ of 18 = ☐

1 mark

18 10 ÷ 10 = ☐

1 mark

© HarperCollins*Publishers* Ltd 2019

19 37 + 43 = ☐

1 mark

20 9 × 3 = ☐

1 mark

© HarperCollins*Publishers* Ltd 2019

21 36 + ☐ = 62

1 mark

22 60 ÷ ☐ = 10

1 mark

© HarperCollins*Publishers* Ltd 2019

23 ☐ + 48 = 63

1 mark

24 $\frac{3}{4}$ of 32 = ☐

1 mark

© HarperCollinsPublishers Ltd 2019

25 $\square - 16 = 62$

1 mark

Total marks ..../25

© HarperCollins*Publishers* Ltd 2019

1 On these cards, the word should match the number.
One of these cards is wrong.
Draw a cross (✗) on the card that is wrong.

1 mark

2 Write each number in the correct place in the diagram.
One has been done for you.

7 24 ~~63~~ 88 109

	Even	Not even
Two-digit number		63
Not two-digit number		

1 mark

3 Write **two different** numbers to make this calculation correct.

☐ + ☐ = 17

1 mark

© HarperCollins*Publishers* Ltd 2019

 Name ____________

4 There are **16** birds in the tree.
9 birds fly away.
How many birds are left?

[ ] birds

1 mark

5 Write the **same** number in **both** boxes to make this number sentence correct.

[ ] × [ ] = 100

1 mark

6 Sam's puppy was **25** centimetres tall.
The puppy grew **8** centimetres in one month.
How tall is the puppy now?

[ ] cm

1 mark

© HarperCollins*Publishers* Ltd 2019

7 Tick (✓) the clock that shows quarter past 12.

1 mark

8 Benito and Ava collect pencil sharpeners.
Benito has collected **27** pencil sharpeners.
Ava has collected **15 more** pencil sharpeners than Benito has.
How many pencil sharpeners has **Ava** collected?

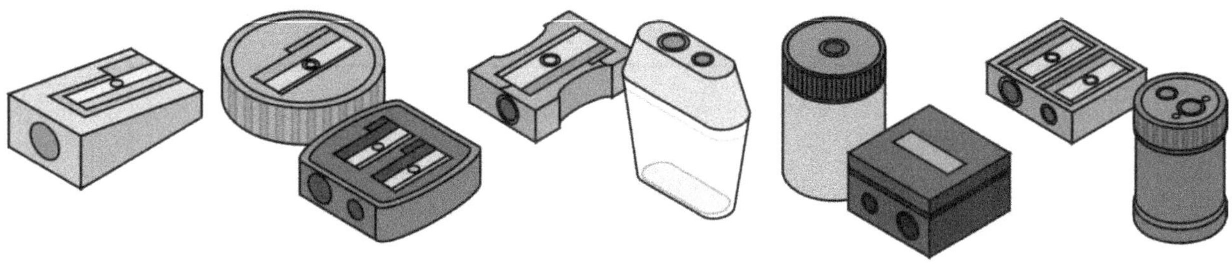

[ ] pencil sharpeners

1 mark

9 Sumi printed a pattern.
She used **two different** faces from the same 3D shape.

a) Tick (✓) the 3D shape Sumi used.

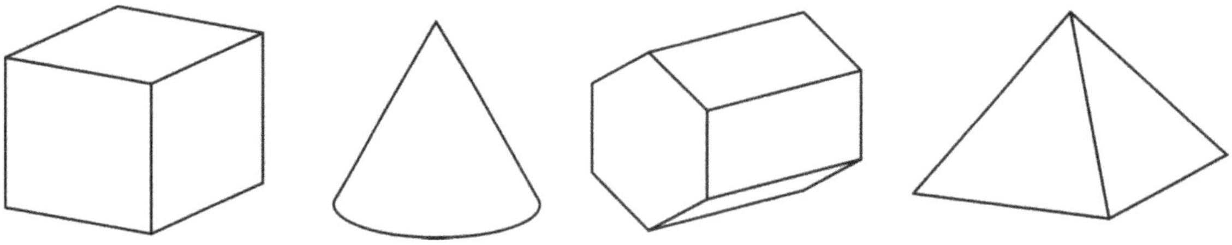

b) Tick the next shape in Sumi's pattern.

2 marks

© HarperCollins*Publishers* Ltd 2019

10 What **fraction** of the square has been shaded?
Circle it.

$\frac{1}{4}$ $\frac{1}{3}$ $\frac{2}{4}$ $\frac{2}{3}$

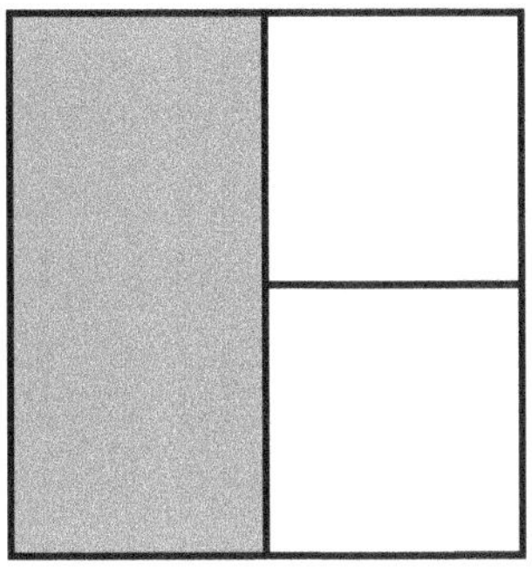

1 mark

11 A shopkeeper has **30** sunflowers and **5** vases.
He puts the same number of sunflowers in each vase.
How many sunflowers go in each vase?

[ ] sunflowers

1 mark

12 Tomas has **20p** in 5p coins.
How many 5p coins does Tomas have?

[ ] coins

1 mark

13 One boat arrives at the dockyard each hour.
How many boats arrive in **1 day**?

[ ] boats

1 mark

© HarperCollins*Publishers* Ltd 2019

14 The bird turns **2 right angles** anticlockwise.
It walks **forward 2 squares**.
Then it turns **1 right angle** anticlockwise.
It walks **forward 1 square** and eats the worm.
Draw a ring around the worm the bird eats.

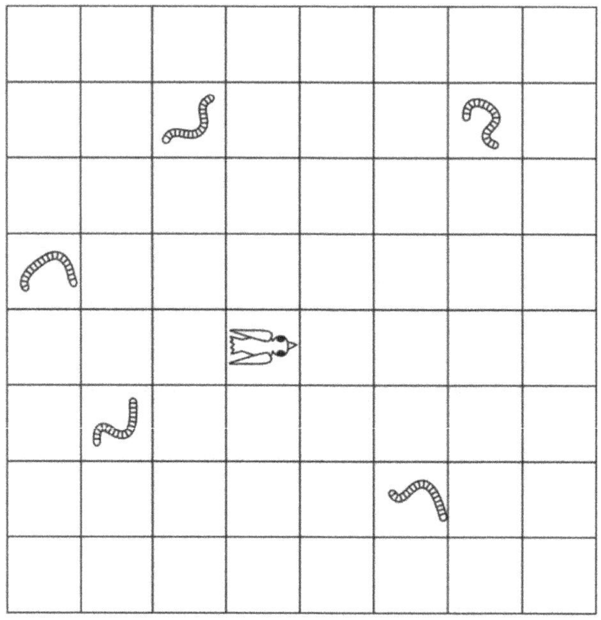

1 mark

15 Children in Year 2 chose their favourite fruit.

Apple	☺☺☺☺
Pear	☺☺☺☺☺☺☺
Banana	☺☺☺☺
Strawberry	☺☺☺☺☺☺☺☺☺☺☺
Mango	☺☺

☺ = 2 children

a) How many children prefer **strawberry**?

[ ] children

b) Which **two** fruits received the same number of votes?

[ ]

2 marks

16

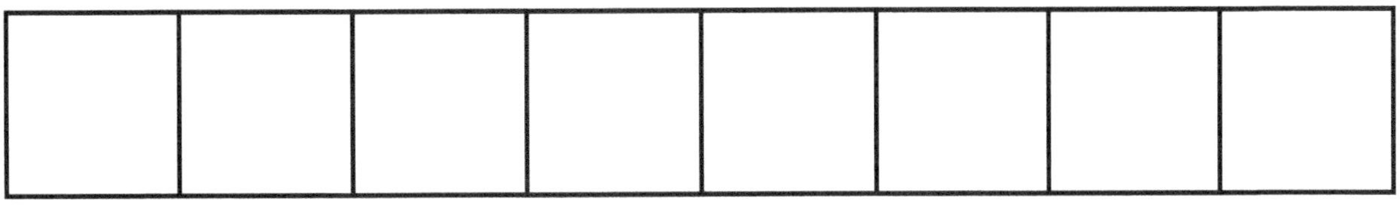

Shade $\frac{3}{4}$ of the strip.

1 mark

17 What time does the clock show?

Tick the correct box.

quarter past 10	ten to 3
10 past 3	quarter to 10

1 mark

18 There are **32** toy cars for the playmat.
27 cars are being played with.
How many cars are **left in the box**?

cars

1 mark

19 Complete the fraction statements.

$\frac{2}{4}$ of £20 = £ ☐

$\frac{3}{4}$ of £20 = £ ☐

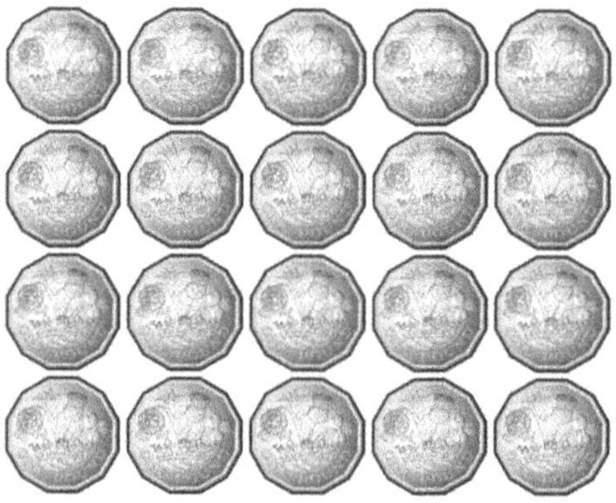

1 mark

20 Complete these calculations.

27 + 5 = ☐ 27 + 50 = ☐

67 – 5 = ☐ 67 – 50 = ☐

1 mark

21 Measure the longest side of the triangle.
Use a ruler.

☐ cm

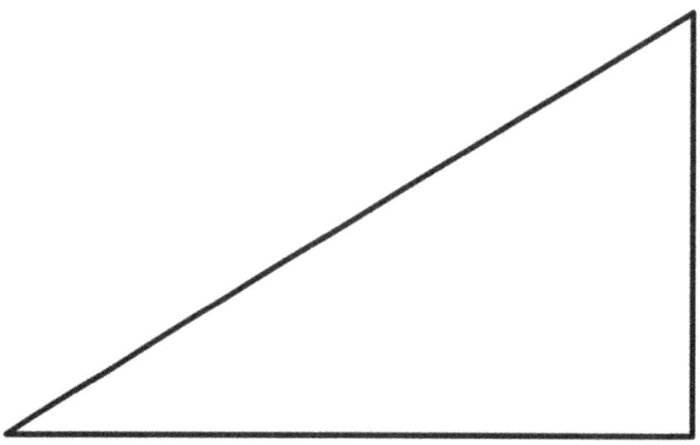

1 mark

22 Lisa buys **1** peach for 32p.
She pays with a **£1** coin.
How much change does Lisa get?

☐ p

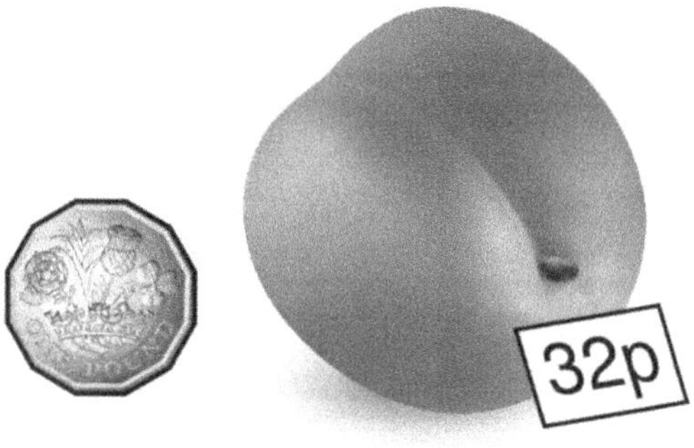

1 mark

© HarperCollins*Publishers* Ltd 2019

23 Sam has **5** packs of stickers.
Write a number sentence that includes × and = to show that Sam has **30** stickers altogether.

1 mark

24 Two of these shapes have **fewer than 9** edges.
Tick (✓) them.

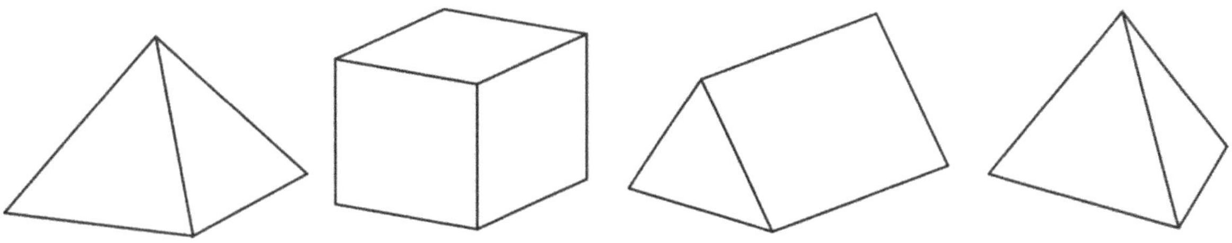

1 mark

25 Look at these numbers.

18	28	38	48

Write each number **once** to make these statements correct.

☐ > ☐

☐ > ☐

1 mark

© HarperCollins*Publishers* Ltd 2019

Year 2 Summer Half Term 1: Reasoning

Name ____________

26 There are **50** colouring pencils in the art set.
Ben takes **12** colouring pencils.
Sumi takes **9** colouring pencils.
How many colouring pencils are left in the art set?

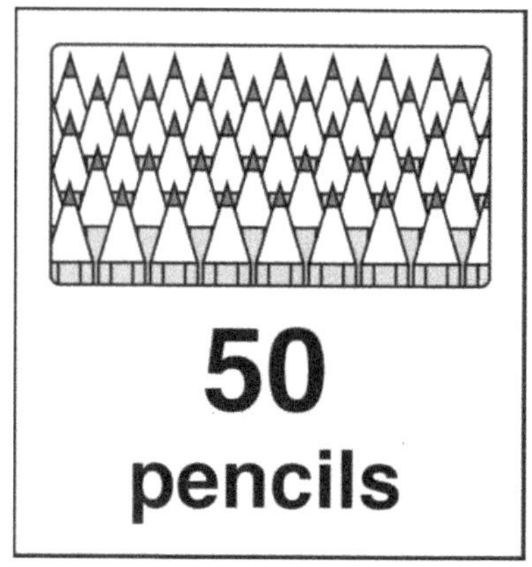

Show your method

[ ] colouring pencils

2 marks

27 There are **24** children in Oak class.
Children work in **pairs (twos)** on an activity.
Complete the number sentence to show how many pairs there are.

[ ] ÷ [ ] = [ ] pairs

1 mark

28 Write the missing number in the box.

$14 + 7 = 10 + \square$

1 mark

© HarperCollins*Publishers* Ltd 2019

29 A shop has **6** bunches of bananas.
Each bunch has **5** bananas.
Mr Chen buys **17** bananas.
How many bananas does
the shop have left?

Show your method

bananas

2 marks

30 $\frac{1}{3}$ of 33p = ☐ p

1 mark

31 Write **two** numbers that are both **greater than 30** to make this subtraction correct.

☐ – ☐ = 9

1 mark

Total marks ………/35

© HarperCollins*Publishers* Ltd 2019

1 9 – 4 = ☐

1 mark

2 13 = 4 + ☐ + 3

1 mark

© HarperCollins*Publishers* Ltd 2019

3 10 + 20 + 30 + 30 = ☐

1 mark

4 46 + 9 = ☐

1 mark

© HarperCollinsPublishers Ltd 2019

5 $7 = 19 - \square$

1 mark

6 $12 \times 2 = \square$

1 mark

© HarperCollinsPublishers Ltd 2019

7 $72 - 7 = \square$

1 mark

8 $\frac{1}{2}$ of $24 = \square$

1 mark

© HarperCollinsPublishers Ltd 2019

9 100 – 60 = ☐

1 mark

10 27 + 48 = ☐

1 mark

© HarperCollinsPublishers Ltd 2019

11 $7 \times 3 =$ ☐

1 mark

12 $54 - 9 =$ ☐

1 mark

© HarperCollinsPublishers Ltd 2019

13 $74 + 26 = \square$

1 mark

14 $5 \div 5 = \square$

1 mark

© HarperCollins*Publishers* Ltd 2019

15 8 + 54 = ☐

1 mark

16 93 + 8 = ☐

1 mark

© HarperCollinsPublishers Ltd 2019

17 $60 \div 5 = \square$

1 mark

18 $49 + \square = 78$

1 mark

© HarperCollins*Publishers* Ltd 2019

19 $\frac{1}{4}$ of 20 = ☐

1 mark

20 83 – 47 = ☐

1 mark

© HarperCollins*Publishers* Ltd 2019

21 $10 \times \square = 70$

1 mark

22 $30 + \square = 83$

1 mark

© HarperCollins*Publishers* Ltd 2019

23 $\frac{3}{4}$ of 28 = ☐

1 mark

24 ☐ – 27 = 56

1 mark

© HarperCollinsPublishers Ltd 2019

25 80 ÷ ☐ = 8

1 mark

Total marks ………/25

© HarperCollins*Publishers* Ltd 2019

 Name ____________

1 Match the words to the numbers.
One has been done for you.

forty-nine	11
one hundred and one	101
ninety-four	49
eleven	94

1 mark

2 Write these numbers in order, starting with the **largest**.

56 43 19 81 34

☐ ☐ ☐ ☐ ☐

largest smallest

1 mark

3 Count in tens.
Start at **37**.
Write the next three numbers in the count.

☐ ☐ ☐

1 mark

4 Measure the length of **one** side of the star.
Use a ruler.

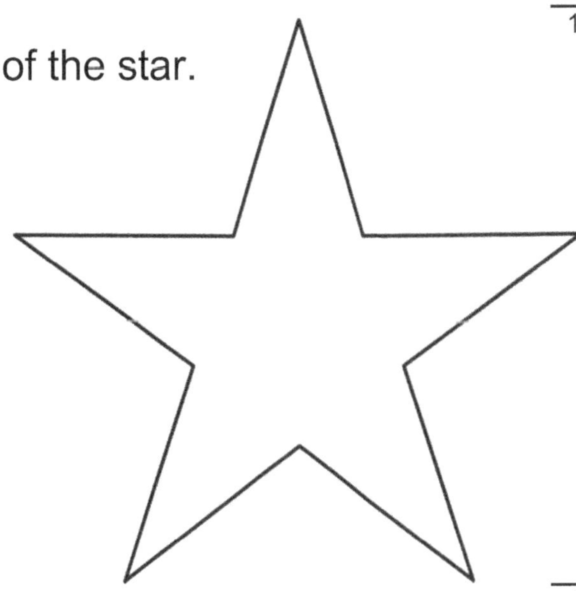

☐ cm

1 mark

© HarperCollinsPublishers Ltd 2019

5 Tick (✓) the **two** statements that are correct. Tick **two**

A pentagon has 5 sides. ☐

A pentagon has curved sides. ☐

A regular pentagon has no lines of symmetry. ☐

A regular pentagon has sides of equal length. ☐

1 mark

6 Jar **A** is full.
It holds **100** marbles.
Estimate how full Jar **B** is.
Tick the matching fraction.

$\frac{1}{2}$ ☐ $\frac{1}{4}$ ☐

$\frac{1}{3}$ ☐ $\frac{3}{4}$ ☐

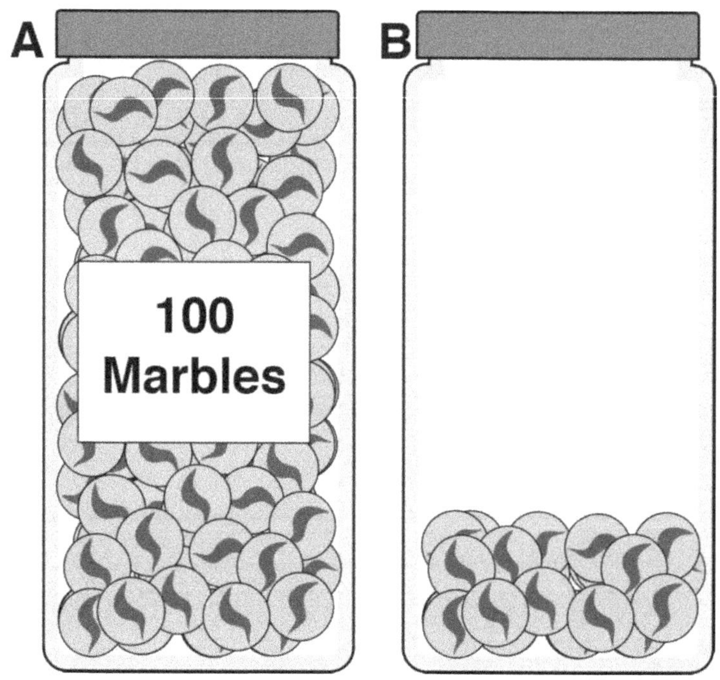

1 mark

7 How much does it cost to buy the **sunglasses** and the **straw hat**?

£ ______

1 mark

 Name ____________

8

Bridgetown School Clubs All clubs start at 4 p.m.		
Monday	Drama	1 hour
Tuesday	Football	$1\frac{1}{2}$ hours
Wednesday	Hockey	1 hour
Thursday	Singing	45 minutes
Friday	Swimming	1 hour

The table lists the different after-school clubs.

a) How many minutes does the **longest** club last?

[minutes]

b) Draw the hands on the clock to show the time when the singing club finishes.

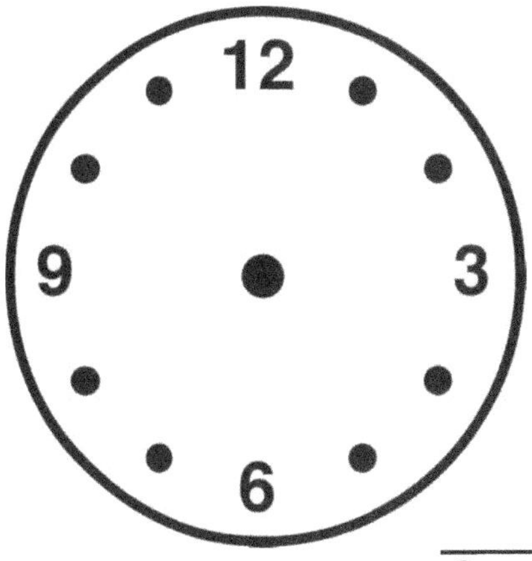

2 marks

9 Look at the thermometer.
It is a hot sunny day.
What is the temperature today?

[°C]

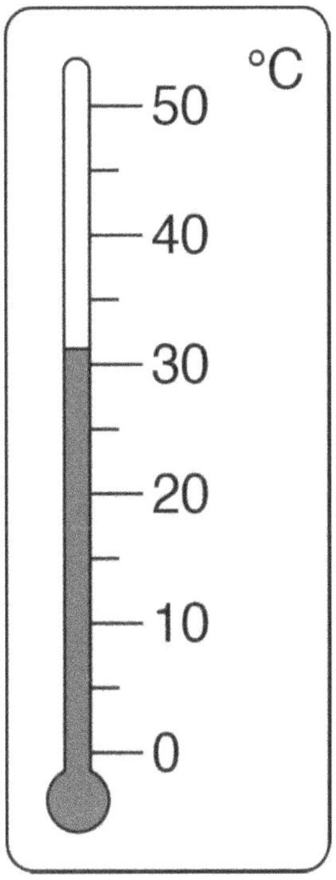

1 mark

© HarperCollins*Publishers* Ltd 2019

10 A dining hall has **8** tables.
Each table has **10** children sitting at it.
Complete the number sentence to show how many children there are **altogether**.

☐ × ☐ = ☐ children

1 mark

11 Join each everyday object to the 3D shape it is most like.
One has been done for you.

Tin of beans	sphere
Box of cereal	cone
Ball	cuboid
Ice-cream cornet	cylinder

(Box of cereal is joined to cuboid.)

1 mark

12 How much **more** water is there in Jug B than in Jug A?

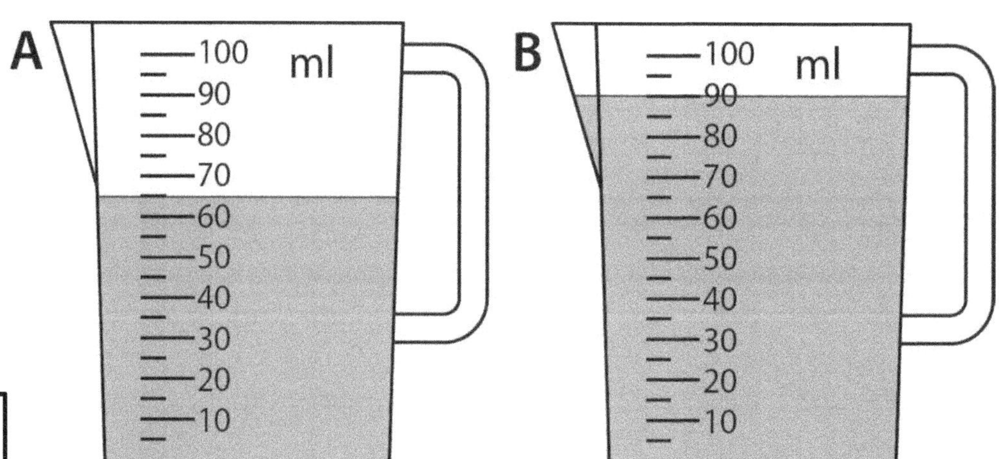

☐ ml

1 mark

© HarperCollins*Publishers* Ltd 2019

13 The bar chart shows the numbers of shells four children collected.

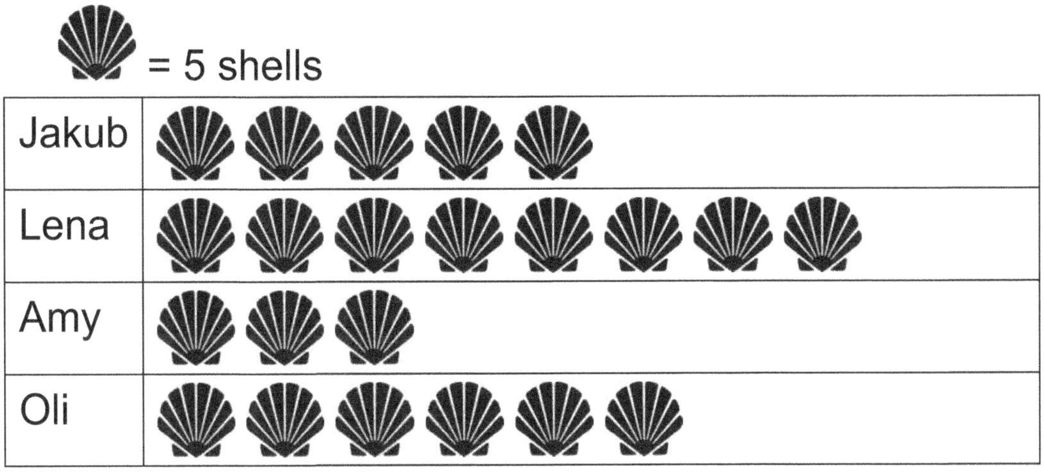

Amy collected fewer shells than Lena.
How many **fewer**?

[ ] shells

1 mark

14 Tomas buys **1** pear for **28p**.
He pays with a **£1 coin**.
How much change does Tomas get?

[ ] p

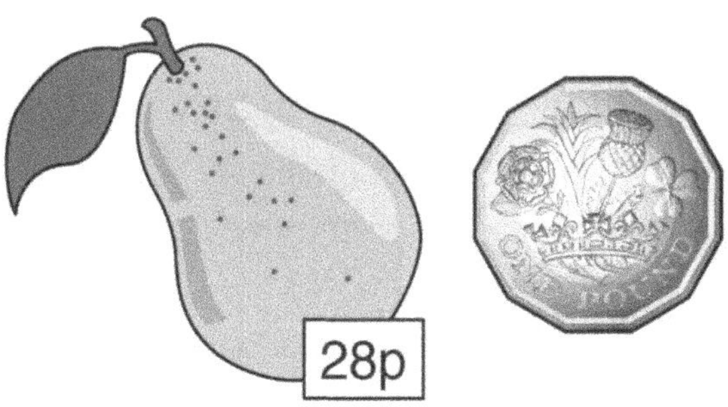

1 mark

15 Emily had **37** marbles.
She won **6 more** marbles in a game.
How many marbles does Emily have now?

[ ] marbles

1 mark

© HarperCollins*Publishers* Ltd 2019

16

Pablo	Patricia	Pamela	Percy
2 kg 100 g	3 kg 20 g	2 kg 10 g	3 kg 200 g

Smallville Zoo has **4** penguins.
Compare the **masses** of the penguins.
Write <, = or > in each box.

Pablo's mass ☐ Percy's mass

Patricia's mass ☐ Pamela's mass

1 mark

17 Use **four different** number cards to complete each of the number sentences below.

10	20	30	40	50	60	70

☐ + ☐ = 80

☐ + ☐ = 80

2 marks

18 Evita has 75p.
Lucy has 20p **less** than Evita.
Tick (✓) the coins **Lucy** has.

1 mark

© HarperCollins*Publishers* Ltd 2019

Year 2 Summer Half Term 2: Reasoning   Name ____________

19 Measure this line.

____

Draw a line that is $\frac{3}{4}$ of the length of this line.

How long is your line?

[cm]

2 marks

20 Write **two** number sentences to match this array.

* * *
* * *
* * *
* * *
* * *
* * *

[ ] × [ ] = [ ]

[ ] ÷ [ ] = [ ]

1 mark

21 Look at the clock.

Tick the clock that shows the time 45 minutes **earlier**.

1 mark

© HarperCollins*Publishers* Ltd 2019

22 The Smith family are going on holiday.
They have **2** bags.
How much do the family's bags weigh altogether?

[kg]

1 mark

23 This turtle has been rotated **clockwise**.

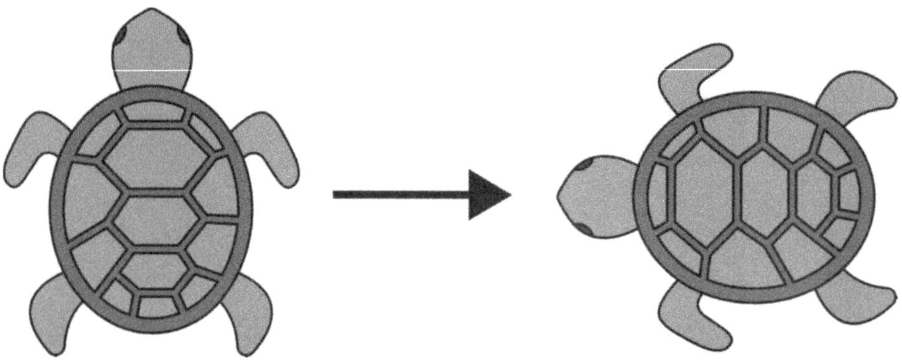

By how much has the turtle been rotated?
Circle your answer.

quarter turn	half turn	three-quarter turn	full turn

1 mark

24 An apple weighs **100 g.**
5 strawberries together weigh the same as the apple.
How much does **one** strawberry weigh?

[g]

1 mark

© HarperCollins*Publishers* Ltd 2019

25 Complete the fraction statement.

$\frac{\square}{\square}$ of 44 = 33

1 mark

26 Sam's swimming lesson **starts** at 20 to 4.
The lesson lasts for **45 minutes**.
Draw the hands on the clock to show the time the lesson finishes.

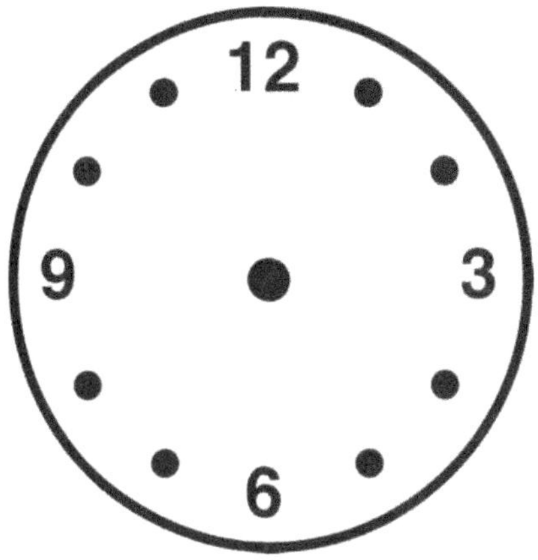

1 mark

27 Tomas, Ella and Clara share the money equally.
How much money do they get each?

£ ____________

1 mark

© HarperCollinsPublishers Ltd 2019

28 There are **8** packets of biscuits.
There are **10** biscuits in each packet.
27 biscuits are eaten at tea time.
How many biscuits are **left**?

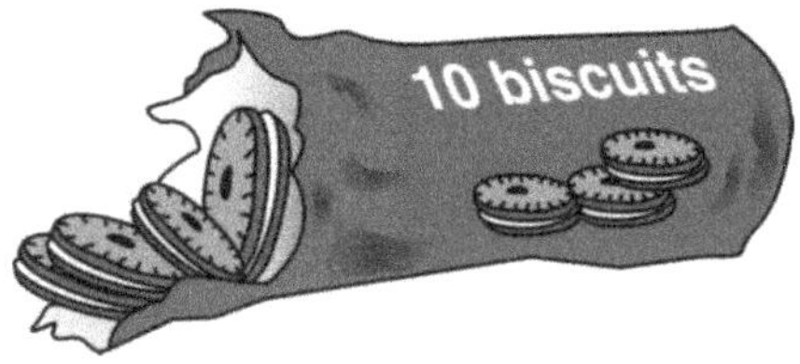

Show your method

□ biscuits

2 marks

29 Write two numbers that are both **greater than 50** to make this subtraction correct.

□ – □ = 7

1 mark

30 Write the missing number in the box.

50 + □ = 48 + 7

1 mark

31 Look at these numbers.

21	22	23	24

Write each number **once** to make these statements correct.

□ < □   □ > □

1 mark

Total marks ..../35

© HarperCollins*Publishers* Ltd 2019

Year 2 Autumn Half Term 1: Arithmetic Mark Scheme

Question	Requirement	Mark	Level of demand
1	40	1	T
2	30	1	T
3	4	1	T
4	90	1	E
5	10	1	E
6	44	1	E
7	5	1	T
8	48	1	E
9	7	1	E
10	50	1	E
11	102	1	E
12	57	1	E
13	8	1	G
14	15	1	G
15	60	1	G

Threshold scores
Working towards the expected standard: 8 or fewer
Working at the expected standard: 9–11
Working at greater depth: 12–15

Balance of difficulty of questions in the paper
4 marks at working towards (T)
8 marks at the expected standard (E)
3 marks at working at greater depth (G)

Year 2 Autumn Half Term 1: Reasoning Mark Scheme

Question	Requirement	Mark	Additional guidance	Level of demand
1	17 (cookies)	1		T
2	76	1		T
3	Second representation (b) ticked. 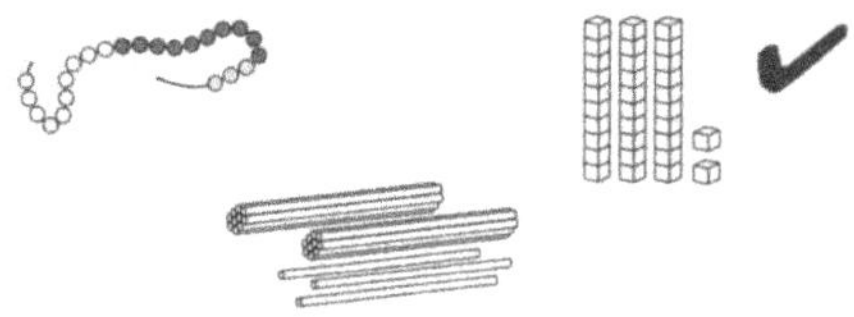	1	Accept any other clear way of indicating the correct response. Do not award the mark if other representations are indicated unless it is clear that the correct representation is the child's final choice.	T
4a	72 written in the first box as shown: 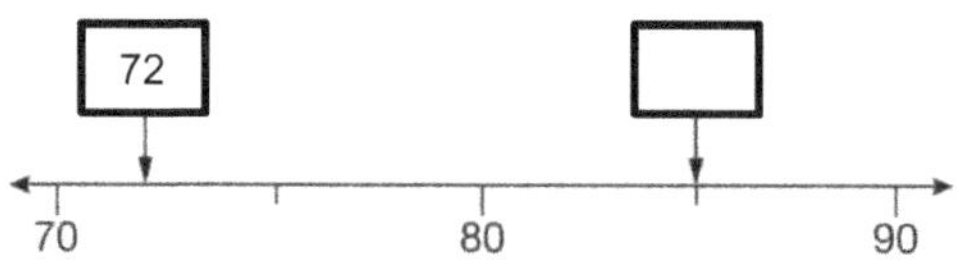 	1	Accept any number in the range 71 to 73.	E
4b	85 written in the second box as shown: 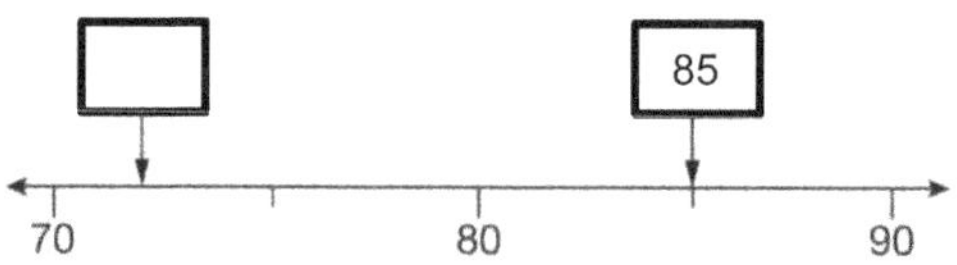 	1		T
5	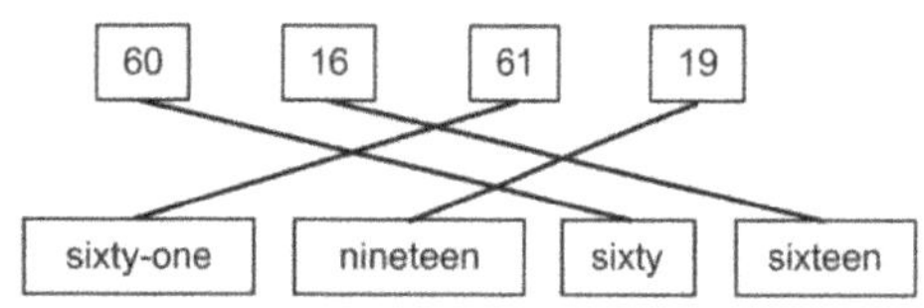 	1	All four pairs must be correctly matched for the award of the mark. Do not award the mark if any numeral or word is matched to more than one numeral or word.	E
6	35	1		E
7	87	1		T
8	Numbers put in the correct order as shown: 29 | 42 | 49 | 92 | 94 smallest … largest	1	All numbers must be in the correct order for the award of the mark. Accept any other clear way of indicating the correct answer, e.g. matching each number to its correct position. Accept the numbers given in the reverse order, provided the labels have been swapped.	E
9	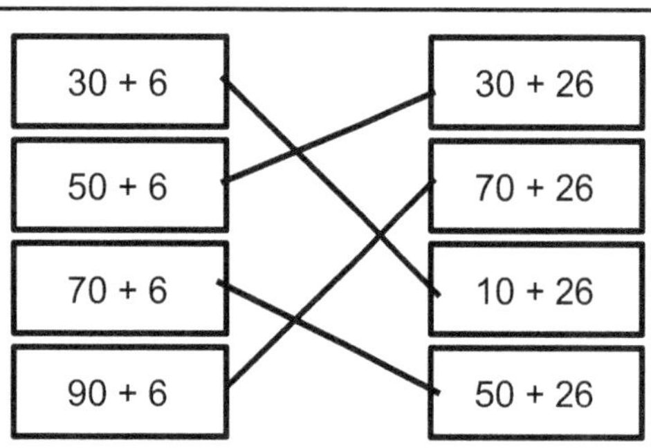 	1	All three pairs must be correctly matched for the award of the mark. Do not award the mark if a sum is matched to more than one sum. Ignore any extra lines drawn from 30 + 6.	G
10	Both numbers correct as shown: 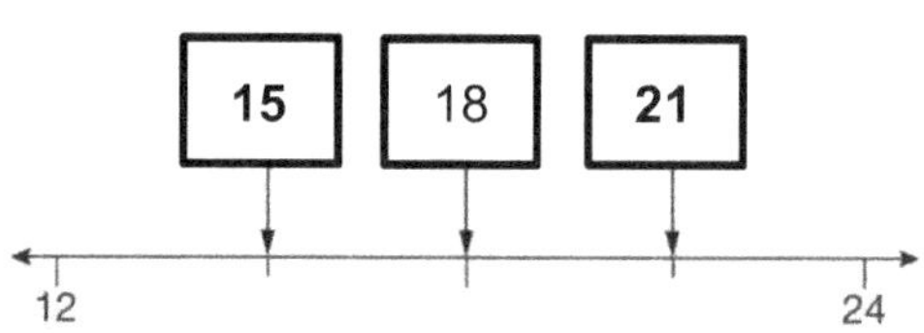 	1	If the answer boxes are empty, accept the correct values written in the correct order elsewhere on the page.	E
11	35 (marbles)	1		G

© HarperCollins*Publishers* Ltd 2019

Year 2 Autumn Half Term 1: Reasoning Mark Scheme

12	20 in upper circle, 49 in lower circle. 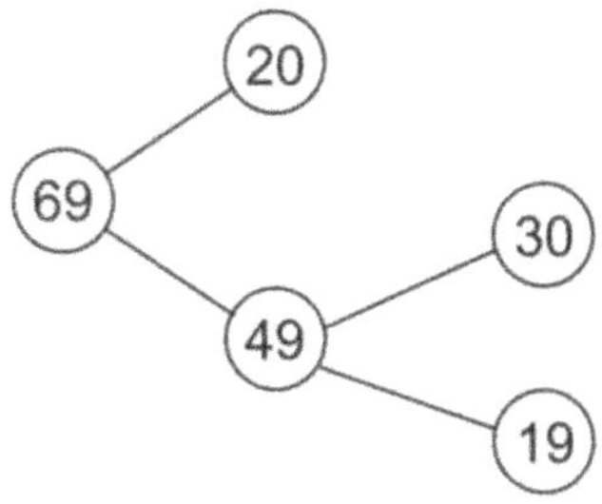	1	Both numbers must be correct for the award of the mark.	G
13	Both inequalities completed correctly, using each of the given numbers once only, i.e. [24] < [37] [37] < [52] [71] > [52] [71] > [24] [24] < [52] [37] < [71] [71] > [37] [52] > [24] [24] < [71] [52] < [71] [52] > [37] [37] > [24]	1	Both inequalities must be correct for the award of the mark. Do not award the mark if any number is used more than once. Do not award the mark if numbers not given in the question are used.	G
14	42	1		E
15	Award 2 marks for the correct answer of 50 pencils. If the answer is incorrect or missing, award ONE mark for evidence of a complete, correct method, e.g. • 100 – 30 – 20 = (incorrect or no answer) • 30 + 20 = 60 (error) 100 – 60 =	2 Or 1		G
16	Calculations completed correctly as shown: [100] – [90] = [10] [100] – [10] = [90]	1	Both calculations must be completed correctly for the award of the mark.	E
17	Both number sentences completed as shown: (number sentences in any pair in any order): [51] – [10] = [41] [10] = [51] – [41] Or [51] – [41] = [10] [41] = [51] – [10]	1	All numbers in both number sentences must be correct for the award of the mark.	E
18	Number sentence completed as shown 43 + 10 = [63] – 10	1		G

Threshold scores
Working towards the expected standard: 11 or fewer
Working at the expected standard: 12–15
Working at greater depth: 16–20

Balance of difficulty of questions in the paper
5 marks at working towards (T)
8 marks at the expected standard (E)
7 marks at working at greater depth (G)

© HarperCollins*Publishers* Ltd 2019

Year 2 Autumn Half Term 2: Arithmetic Mark Scheme

Question	Requirement	Mark	Level of demand
1	90	1	T
2	79	1	T
3	6	1	T
4	35	1	T
5	58	1	E
6	7	1	E
7	75	1	E
8	37	1	E
9	10	1	G
10	22	1	E
11	9	1	E
12	43	1	G
13	12	1	G
14	27	1	E
15	8	1	G

Threshold scores
Working towards the expected standard: 8 or fewer
Working at the expected standard: 9–11
Working at greater depth: 12–15

Balance of difficulty of questions in the paper
4 marks at working towards (T)
7 marks at the expected standard (E)
4 marks at working at greater depth (G)

© HarperCollins*Publishers* Ltd 2019

Year 2 Autumn Half Term 2: Reasoning Mark Scheme

Question	Requirement	Mark	Additional guidance	Level of demand
1	Both correct numbers circled as shown. (57) 62 (91) 34	1	Both numbers must be indicated for the award of the mark. Accept any other clear way of indicating the two correct numbers. **Do not** award the mark if additional numbers are indicated, unless it is clear that the two correct numbers are the child's final choice.	T
2	70 (crayons)	1		T
3	8, 9	1	Both numbers must be recorded for the award of the mark. Accept the numbers in any order.	E
4	3 + 7 = [10] 13 + 7 = [20] 30 + 70 = [100]	1	All three sums must be correct for the award of the mark.	T
5	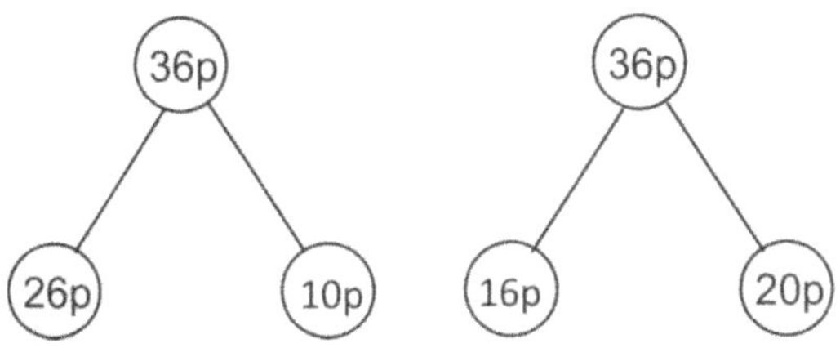	1	Both part–whole models must be correctly completed for the award of the mark.	T
6	45 (cupcakes)	1		T
7	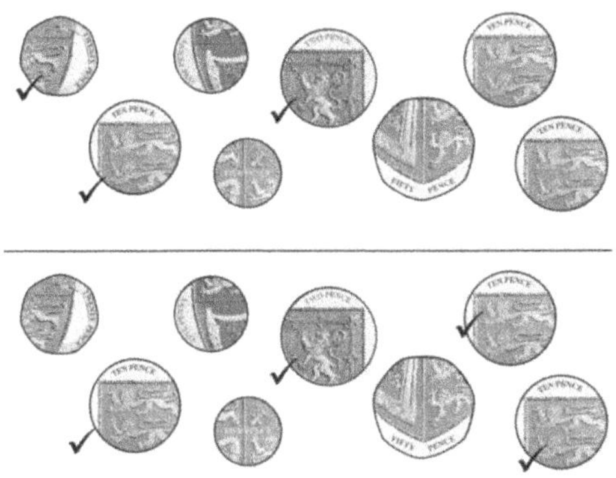 **Either** 20p, 10p, 2p ticked **or** 10p, 10p, 10p, 2p ticked.	1	Accept any other clear way of indicating a correct response. **Do not** award the mark if other responses are indicated unless it is clear that a correct response is the child's final choice.	E
8	The correct calculation ticked as shown. [5 + 4] [5 × 4] ✓ [5 – 4] [5 + 5 + 5]	1	Accept any other clear way of indicating a correct response. **Do not** award the mark if other responses are indicated unless it is clear that a correct response is the child's final choice.	E
9	9 written in both boxes as shown. [9] + [9] = [18]	1	The correct number must be written in both boxes for the award of the mark.	E

© HarperCollins*Publishers* Ltd 2019

Year 2 Autumn Half Term 2: Reasoning Mark Scheme

10	The three sums completed correctly using six different numbers, e.g. 10 + 41 = 51 20 + 31 = 51 30 + 21 = 51	2	All three sums must be correct for the award of **2 marks**. Accept 0 + 51 as a correct answer. Award **1 mark** for any **two** sums completed correctly, such that all three calculations are correct but numbers are repeated in two of the calculations or there is an error in one of the calculations, e.g. 20 + 31 = 51 31 + 20 = 51 30 + 21 = 51	E
11	Both calculations circled as shown. 10 + 6 (6 × 10) 6 + 6 + 6 + 6 + 6 + 6 (10 + 10 + 10 + 10 + 10 + 10)	1	Both correct calculations must be indicated for the award of the mark. Accept any other clear way of indicating the correct calculations, including evaluating only the correct two calculations, i.e. writing 60 alongside each of the two correct calculations. Do not award the mark if other calculations have been evaluated, and the correct two have not been indicated. Do not award the mark if more than two calculations are circled unless it is clear that the correct calculations are the child's final choice.	E
12	Both numbers correct as shown. 10 9 12 7 5 14	1	Both numbers must be correct for the award of the mark.	E
13	Both signs written correctly as shown. $10 \times 5 > 12 \times 2$ $10 \times 2 < 5 \times 5$	1	Both signs must be correct for the award of the mark. Accept slight inaccuracies in the drawing of the signs, as long as the intention is clear.	G
14	48 (stickers)	1		E
15	Both statements correctly ticked or crossed. 54 < 84 – 30 × 67 > 70 + 7 × 85 = 90 – 5 ✓	1	Both statements must be correctly marked for the award of the mark.	E
16	£40	1		E
17	25 (balloons)	1		E
18	10 (coins)	1		E
19	35 (p)	1		E
20	Both multiplication calculations completed. 10 × 3 = 30 3 × 10 = 30	1	Both number sentences must be correctly completed for the award of the mark. Accept the completed calculations in any order.	G

© HarperCollins*Publishers* Ltd 2019

Year 2 Autumn Half Term 2: Reasoning Mark Scheme

21	90 (p)	2	Award **2 marks** for the correct answer of 90 (p). If the answer is incorrect or missing, award **1 mark** for evidence of a complete, correct method, e.g. 15 + 15 + 20 + 20 + 20 = (incorrect or no answer) 15 x 2 = 20 (error) 20 + 60 =	G
22	Both multiplication calculations completed: [7] × [5] = 35 or [5] × [7] = 35 **and** one of the following. [10] × [4] = 40 [4] × [10] = 40 [5] × [8] = 40 [8] × [5] = 40	1	Both number sentences must be correctly completed for the award of the mark.	G
23	Calculation completed correctly as shown. 27 + 7 − [15] = 19	1		G

All numbers in both number sentences must be correct for the award of the mark.

Threshold scores
Working towards the expected standard: 14 or fewer
Working at the expected standard: 15–19
Working at greater depth: 20–25

Balance of difficulty of questions in the paper
5 marks at working towards (T)
14 marks at the expected standard (E)
6 marks at working at greater depth (G)

Year 2 Spring Half Term 1: Arithmetic Mark Scheme

Question	Requirement	Mark	Level of demand
1	70	1	T
2	56	1	T
3	90	1	T
4	9	1	E
5	24	1	E
6	15	1	T
7	13	1	T
8	11	1	E
9	45	1	E
10	3	1	T
11	14	1	T
12	100	1	E
13	30	1	E
14	10	1	G
15	35	1	E
16	14	1	G
17	107	1	G
18	23	1	E
19	47	1	G
20	5	1	G

Threshold scores
Working towards the expected standard (T): 11 or fewer
Working at the expected standard (E): 12–15
Working at greater depth (G): 16–20

Balance of difficulty of questions in the paper
7 marks at working towards (T)
8 marks at the expected standard (E)
5 marks at working at greater depth (G)

Year 2 Spring Half Term 1: Reasoning Mark Scheme

Question	Requirement	Mark	Additional guidance	Level of demand
1	10 × 8 = 80 **or** 8 × 10 = 80	1	Accept either calculation written outside of the recording frame.	T
2	4, 13, 22, 31, 40	1	All five numbers must be recorded for the award of the mark. Accept the five numbers in any order.	E
3	triangle ☐ square ✓ rectangle ☐ circle ☐ hexagon ✓	1	Both shapes must be ticked for the award of the mark. Accept any other clear way of indicating a correct response.	T
4	2 (marbles)	1	Accept any clear indication of 2 marbles each if the answer has not been recorded in the answer box.	T
5	17 + 8 — 10 + 15 37 + 8 — 30 + 15 57 + 8 — 50 + 15 77 + 8 — 70 + 15	1	All calculations must be correctly matched. Accept any other clear way of indicating a correct response.	E
6	24 + 24 = 48	1	The correct number must be written in both boxes for the award of the mark.	E
7	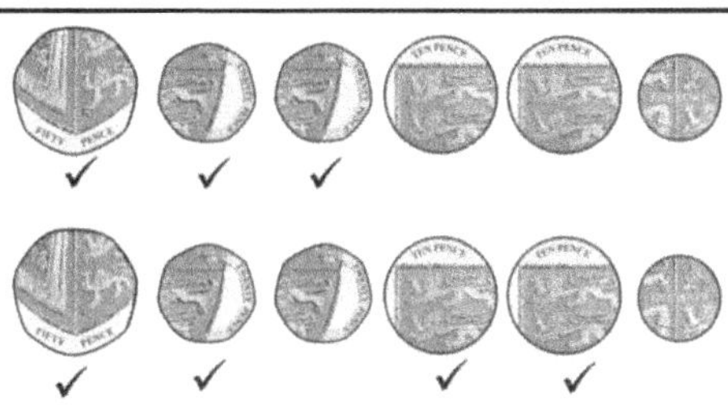	2	Award **2 marks** for both sets of coins correctly marked. Award **1 mark** if one set of coins is correctly marked. Accept any other clear way of indicating a correct response.	T
8	45 ÷ 9 = 5 **or** 45 ÷ 5 = 9 **and** 5 × 9 = 45	1	Both number sentences must be correct for the award of the mark.	E
9	Some possible answers: 10 + 20 + 70 = 100 20 + 20 + 60 = 100 10 + 30 + 60 = 100 10 + 40 + 50 = 100	1	All calculations must be in multiples of 10 and total 100. Accept 0 as a value. **Do not** accept reordered number sentences as different, e.g. 10 + 40 + 50 = 100 and 10 + 50 + 40 = 100 are not different number sentences.	G
10	12 × 2 = 24 children	1	Accept 2 × 12 = 24 (children)	E
11	26 (years)	1		E

Year 2 Spring Half Term 1: Reasoning Mark Scheme

12	25 (p)	1	Accept drawing of coins or other methods of indicating the correct amounts.	G
13a	4 (children)	1		E
13b	27 (children)	1		E
14	Hexagon ticked as indicated. 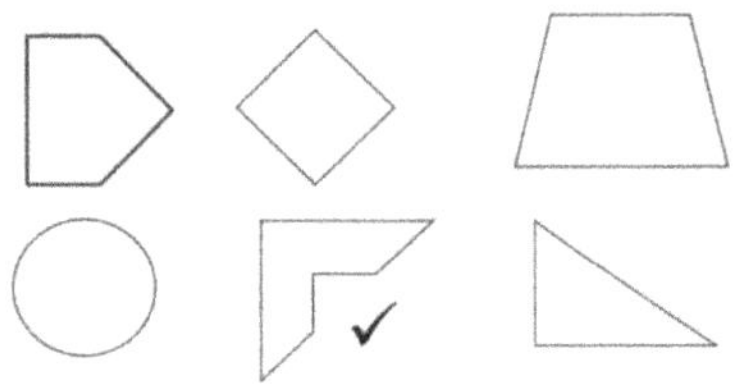	1	Accept any other clear way of indicating the correct shape. **Do not** award the mark if additional shapes are indicated, unless it is clear that the correct shape is the child's final choice.	E
15	(£)29	1		T
16	Calculations completed as below. 74 = 56 + 18 87 > 56 + 17 62 < 56 + 16	1	All three signs must be correct for the award of the mark.	E
17	27 (pens)	1		E
18	**Either** 20p, 10p, 2p ticked **or** 10p, 10p, 10p, 2p ticked.	1	Accept any method of indicating the correct coins.	E
19	Square-based pyramid and triangular prism ticked.	1	Accept any method of indicating the correct shapes.	G
20	Either [10] × [3] = 30 **or** [3] × [10] = 30 And either [6] × [5] = 30 **or** [5] × [6] = 30	1	Accept 2 × 15 = 30 or 15 × 2 = 30 in place of one of the calculations.	G
21	28 (cars)	2	Award **2 marks** for the correct answer of 28 (cars) If the answer is incorrect or missing, award **1** mark for evidence of a complete, correct method, e.g. • 57 + 16 – 45 = (incorrect or no answer) • 57 + 16 = 72 (error) 72 – 45 =	G
22	34 – 18 = 16 ✓ 18 + 16 = 34 18 – 16 = 34 34 – 16 = 18 ✓	1	Do not award the mark if other options are ticked.	G

Threshold scores
Working towards the expected standard (T): 14 or fewer
Working at the expected standard (E): 15–19
Working at greater depth (G): 20–25

Balance of difficulty of questions in the paper
6 marks at working towards (T)
12 marks at the expected standard (E)
7 marks at working at greater depth (G)

© HarperCollins*Publishers* Ltd 2019

Year 2 Spring Half Term 2: Arithmetic Mark Scheme

Question	Requirement	Mark	Level of demand
1	70	1	T
2	30	1	T
3	16	1	T
4	9	1	T
5	19	1	T
6	9	1	E
7	53	1	E
8	63	1	E
9	60	1	T
10	100	1	E
11	5	1	G
12	63	1	E
13	21	1	E
14	7	1	E
15	40	1	E
16	28	1	E
17	3	1	G
18	58	1	G
19	5	1	G
20	9	1	G

Threshold scores
Working towards the expected standard (T): 11 or fewer
Working at the expected standard (E): 12–15
Working at greater depth (G): 16–20

Balance of difficulty of questions in the paper
6 marks at working towards (T)
9 marks at the expected standard (E)
5 marks at working at greater depth (G)

© HarperCollins*Publishers* Ltd 2019

Year 2 Spring Half Term 2: Reasoning Mark Scheme

<table>
<tr><th>Question</th><th>Requirement</th><th>Mark</th><th>Additional guidance</th><th>Level of demand</th></tr>
<tr><td>1</td><td>Both correct numbers circled as shown:
17   (54)
(82)   39</td><td>1</td><td>Both numbers must be indicated for the award of the mark.
Accept any other clear way of indicating the two correct numbers.
Do not award the mark if additional numbers are indicated, unless it is clear that the two correct numbers are the child's final choice.</td><td>T</td></tr>
<tr><td>2</td><td>Letters put in the correct order as shown.
C D A B
longest   shortest</td><td>1</td><td>All letters must be in the correct order for the award of the mark.
Accept any other clear way of indicating the correct answer, e.g. matching each snake to its correct position.</td><td>T</td></tr>
<tr><td>3</td><td>The correct fraction ticked as shown:
$\frac{1}{2}$ ☐ $\frac{1}{4}$ ☐
$\frac{1}{3}$ ☑ $\frac{3}{4}$ ☐</td><td>1</td><td>Accept any other clear way of indicating the correct answer.
Do not award the mark if additional fractions are indicated, unless it is clear the correct fraction is the child's final choice.</td><td>T</td></tr>
<tr><td>4</td><td>Table correctly completed as shown.<table><tr><th>3D shapes</th><th>faces</th><th>vertices</th><th>edges</th></tr><tr><td>cuboid</td><td>6</td><td>8</td><td>12</td></tr><tr><td>square-based pyramid</td><td>5</td><td>5</td><td>8</td></tr></table></td><td>1</td><td></td><td>T</td></tr>
<tr><td>5</td><td>26 written in both boxes as shown.
26 + 26 = 52</td><td>1</td><td>The correct number must be written in both boxes for the award of the mark.</td><td>E</td></tr>
<tr><td>6</td><td>Correct box ticked as shown.
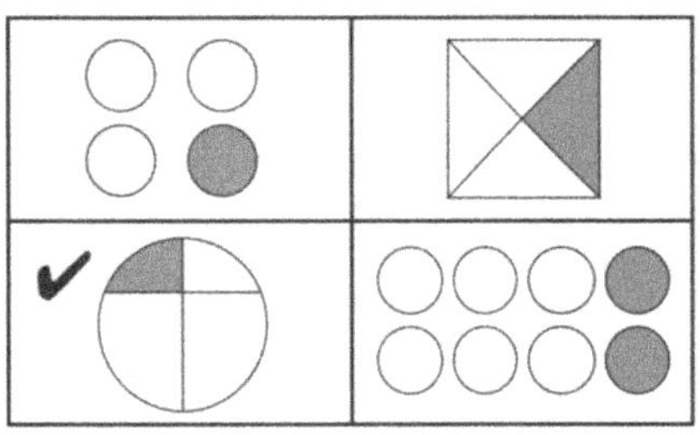</td><td>1</td><td>Accept any other clear way of indicating the correct answer.
Do not award the mark if additional boxes are indicated, unless it is clear the correct box is the child's final choice.</td><td>T</td></tr>
<tr><td>7</td><td>35 (cm)</td><td>1</td><td></td><td>T</td></tr>
<tr><td>8a</td><td>9 (stickers)</td><td>1</td><td></td><td>E</td></tr>
<tr><td>8b</td><td>Sam and Erin.</td><td>1</td><td>Both children must be named for award of the mark. Accept errors in spelling provided the intention is clear.</td><td>E</td></tr>
<tr><td>9</td><td>$\frac{3}{4}$ of the shape shaded, e.g.
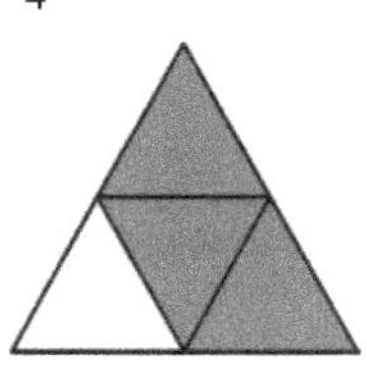</td><td>1</td><td>Accept any other clear way of indicating three-quarters, e.g. marking the appropriate number of sections. Accept slight inaccuracies in shading.
Do not award the mark if more or less than three-quarters has been shaded and the intention is not clear.</td><td>E</td></tr>
</table>

© HarperCollins*Publishers* Ltd 2019

Year 2 Spring Half Term 2: Reasoning Mark Scheme

10	Award **2 marks** for the three sums completed correctly, using six different numbers, e.g. 71 + 1 = 72 70 + 2 = 72 60 + 12 = 72 Award **1 mark** for any two sums completed correctly, such that all three calculations are correct but numbers are repeated in two of the calculations or there is an error in one of the calculations, e.g. 71 + 1 = 72 1 + 72 = 72	2 or 1	All three sums must be correct for the award of **2 marks**. Accept 0 + 72 as a correct answer. Any two sums can be correct for the award of **1 mark**.	E
11	(£)48	1		E
12	61 (bricks)	1		E
13	7 (coloured pencils)	1		E
14	Shape indicated as shown: 	1	Accept any other clear way of indicating the correct answer, e.g. circling or putting a cross on the correct shape. **Do not** award the mark if more than one shape has been indicated, unless it is clear that the correct shape is the child's final choice.	E
15	Object and length match as indicated. 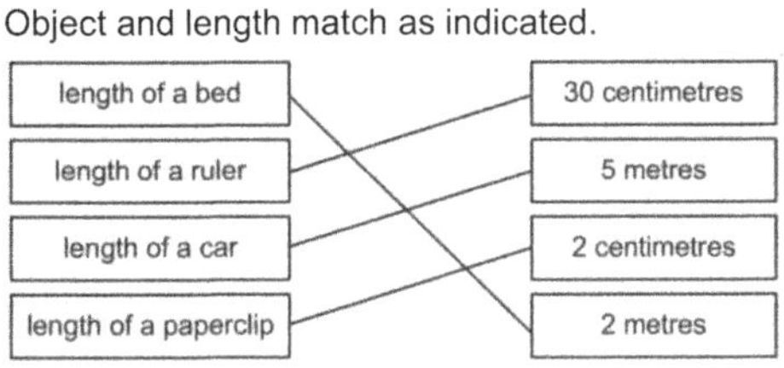	1	All objects and lengths must be correctly matched. Accept any other clear way of indicating a correct response.	E
16	30 (years)	1		E
17	Correct numbers marked as shown: 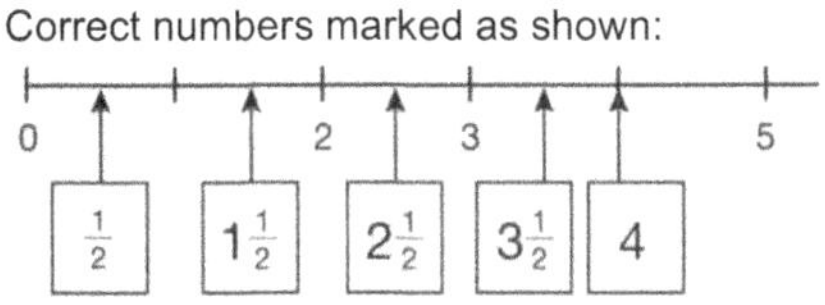	1	All **three** numbers must be correctly marked for the award of the mark. Accept other ways of recording, for example in words: one and a half or 1 and a half.	G
18	10 (coins)	1		E
19	Square drawn with the correct dimensions, e.g. 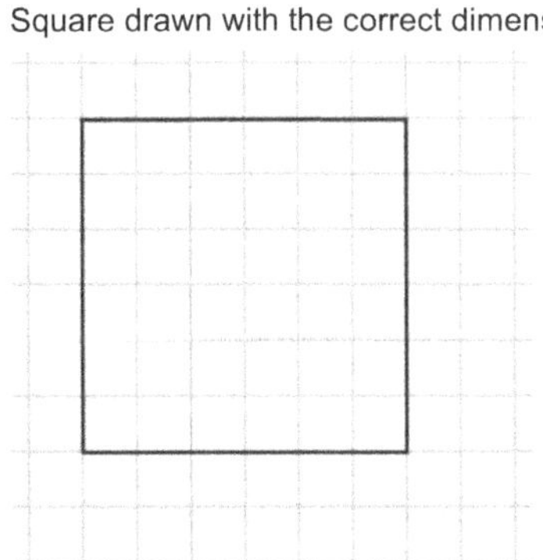	1	Accept any orientation of the square, with the correct dimensions. Accept slight inaccuracies of drawing the square as long as the intention is clear; allowing a tolerance of up to 5 mm.	E

© HarperCollins*Publishers* Ltd 2019

Year 2 Spring Half Term 2: Reasoning Mark Scheme

20	Award **2 marks** for the correct answer of 12 (toy figures). If the answer is incorrect or missing, award **1 mark** for evidence of a partial or complete, correct method, e.g. • 3 + 3 = 6, 6 + 6 = 10 (incorrect or no answer) • 3 + 3 + 3 = 9 (partially correct method).	2 or 1		G
21	Both number sentences completed as shown. 2 × 9 = 18 18 ÷ 2 = 9 or 2 × 9 = 18 18 ÷ 9 = 2	1	All numbers in both number sentences must be correct for the award of the mark.	E
22	Award the mark for any combination of coins indicated that totals 35p, e.g.  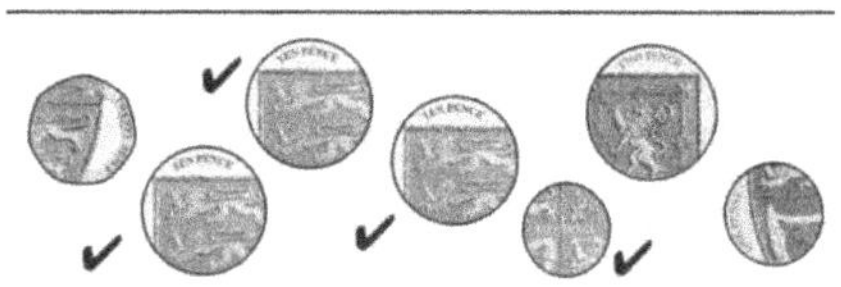	1	Accept any other clear way of indicating a correct combination of coins, i.e. 20p, 10p and 5p or 10p, 10p, 10p and 5p	E
23	53 (centimetres)	1		E
24	Calculations completed as below. 3 metres > 30 centimetres 105 cm < 1 m 15 cm 100 cm = 1 metre	1	All three signs must be correct for the award of the mark.	E
25	96 (football cards)	1		G
26	Calculation completed correctly as shown. 24 + 18 – 30 = 12	1		G
27	Checking calculation as shown. 16 + 29 = 45	1	Accept 29 + 16 = 45	G

Threshold scores
Working towards the expected standard (T): 17 or fewer
Working at the expected standard (E): 18–23
Working at greater depth (G): 24–30

Balance of difficulty of questions in the paper
6 marks at working towards (T)
18 marks at the expected standard (E)
6 marks at working at greater depth (G)

© HarperCollins*Publishers* Ltd 2019

Year 2 Summer Half Term 1: Arithmetic Mark Scheme

Question	Requirement	Mark	Level of demand
1	8	1	T
2	100	1	T
3	43	1	T
4	21	1	T
5	55	1	T
6	32	1	E
7	7	1	E
8	70	1	E
9	13	1	E
10	70	1	T
11	82	1	E
12	7	1	E
13	100	1	E
14	55	1	E
15	11	1	E
16	103	1	E
17	6	1	E
18	1	1	E
19	80	1	E
20	27	1	G
21	26	1	G
22	6	1	G
23	15	1	G
24	24	1	G
25	78	1	G

Threshold scores
Working towards the expected standard: 14 or fewer
Working at the expected standard: 15–19
Working at greater depth: 20–25

Balance of difficulty of questions in the paper
6 marks at working towards (T)
13 marks at the expected standard (E)
6 marks at working at greater depth (G)

© HarperCollins*Publishers* Ltd 2019

Year 2 Summer Half Term 1: Reasoning: Mark Scheme

<table>
<tr><th>Question</th><th>Requirement</th><th>Mark</th><th>Additional guidance</th><th>Level of demand</th></tr>
<tr><td>1</td><td>Card indicated as shown.

</td><td>1</td><td>Accept any other clear way of indicating the correct answer.
Do not award the mark if other cards are indicated unless it is clear the correct card is the child's final choice.</td><td>T</td></tr>
<tr><td>2</td><td>All four numbers added to the diagram as shown.
<table><tr><th></th><th>Even</th><th>Not even</th></tr><tr><td>Two-digit number</td><td>24 88</td><td>63</td></tr><tr><td>Not two-digit number</td><td></td><td>7 109</td></tr></table></td><td>1</td><td>Accept any other clear way of indicating the correct answer.
Do not award the mark if a number is written or indicated in more than one square unless it is clear the correct card is the child's final choice.</td><td>T</td></tr>
<tr><td>3</td><td>Two different numbers written that total 17, e.g.
12 + 5 = 17
14 + 3 = 17
1 + 16 = 17</td><td>1</td><td>Accept any two different numbers that total 17, including 17 + 0 = 17.</td><td>T</td></tr>
<tr><td>4</td><td>7 (birds)</td><td>1</td><td></td><td>T</td></tr>
<tr><td>5</td><td>Number sentence completed correctly as shown.
10 × 10 = 100</td><td>1</td><td>The number 10 must appear in both answer boxes for the award of the mark.</td><td>T</td></tr>
<tr><td>6</td><td>33 (cm)</td><td>1</td><td></td><td>T</td></tr>
<tr><td>7</td><td>The correct time ticked as shown.
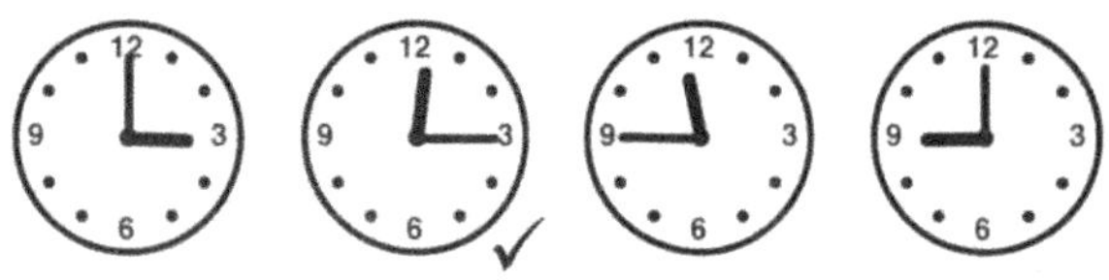</td><td>1</td><td>Accept any other clear way of indicating the correct answer.
Do not award the mark if additional times are indicated, unless it is clear that the correct time is the child's final choice.</td><td>E</td></tr>
<tr><td>8</td><td>42 (pencil sharpeners)</td><td>1</td><td></td><td>E</td></tr>
<tr><td>9</td><td>a)
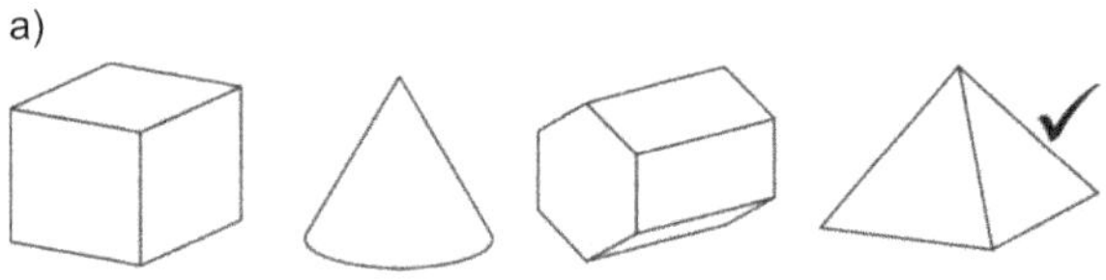
b)
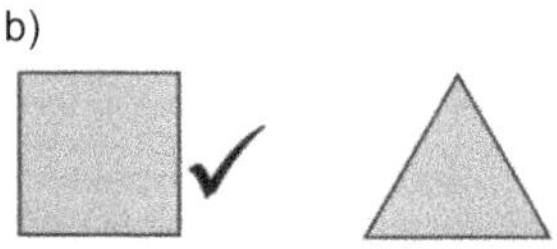</td><td>2 or 1</td><td>Award 1 mark for one correct answer.
a) Accept hexagonal prism changed to triangular prism and ticked since a triangular prism could have square faces.</td><td>a) G
b) E</td></tr>
<tr><td>10</td><td>Correct fraction circled as shown.
$\frac{1}{4}$ $\frac{1}{3}$ ($\frac{2}{4}$ circled) $\frac{2}{3}$</td><td>1</td><td>Accept any other clear way of indicating the correct answer.
Do not award the mark if additional fractions are indicated, unless it is clear the correct fraction is the pupil's final choice.
Do not accept alternative equivalent values written, e.g. the word 'half'.</td><td>E</td></tr>
</table>

Year 2 Summer Half Term 1: Reasoning: Mark Scheme

11	6 (sunflowers)	1	The correct number of 6 must be seen for the award of the mark. Also accept the word 'six' written as long as it is clear it is the child's final answer.	T
12	4 (coins)	1		T
13	24 (boats)	1		E
14	Correct worm circled as shown. 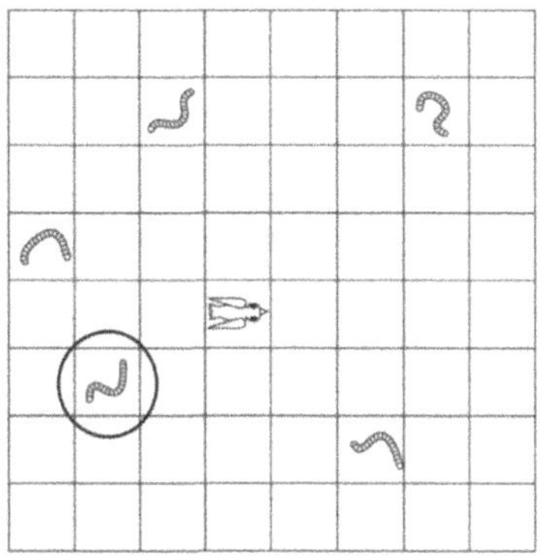	1	Accept any other clear way of indicating the correct answer. Do not award the mark if additional worms are indicated, unless it is clear that the correct worm is the child's final choice.	E
15	a) 22 (children) b) apple and banana	1 1	Award **1 mark** for each correct answer. b) both fruits must be stated for the award of the mark.	a) E b) E
16	Any 6 of the 8 squares shaded, e.g. 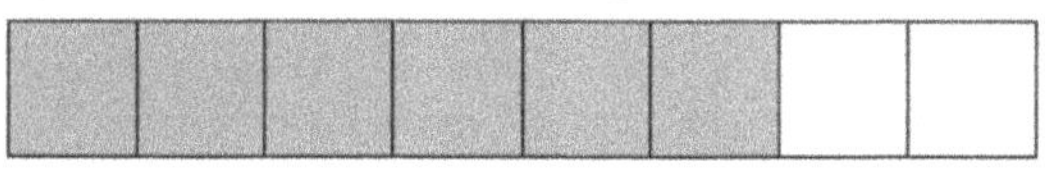	1	Accept any other clear way of indicating the correct fraction; e.g. marking the appropriate number of sections. Do not award the mark if additional shading is present, unless it is clear that the correct shading is the child's final choice.	E
17	The correct time ticked as shown. quarter past 10	ten to 3 ✓ 10 past 3	quarter to 10	1	Accept any other clear way of indicating the correct answer. Do not award the mark if additional times are indicated, unless it is clear that the correct time is the child's final choice.	E
18	5 (cars)	1		E
19	Fraction statements completed as shown. $\frac{2}{4}$ of £20 = £10 $\frac{3}{4}$ of £20 = £15	1	Both statements must be correctly completed for the award of the mark.	E
20	Calculations completed as indicated. 27 + 5 = 32	27 + 50 = 77 67 – 5 = 62	67 – 50 = 17	1	All **four** calculations must be correctly completed for the award of the mark.	E
21	10 (cm)	1	Accept any number **in the** range $9\frac{1}{2}$ to $10\frac{1}{2}$ **inclusive,** including decimal equivalents.	E

© HarperCollins*Publishers* Ltd 2019

22	68(p)	1		E
23	A correct number sentence given, e.g. 5 × 6 = 30 6 × 5 = 30	1	Award the mark even if additional correct or relevant calculations are given along with a correct calculation, e.g. 2 × 5 = 10 3 × 5 = 15 4 × 5 = 20 5 × 5 = 25 Also accept: 5 × 6 = 1 × 30 5 + 5 + 5 + 5 + 5 + 5 = 1 × 30 **Do not** accept 1 × 30 or 30 × 1 unless accompanied by an additional correct number sentence. **Do not** accept an incomplete number sentence e.g. 6 × 5 6 × 5 30 (missing equals sign) 6 × 4 = (missing product)	E
24	Both correct shapes ticked as shown. 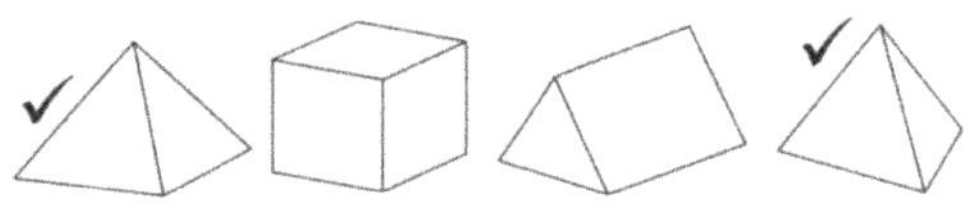	1	Accept any other clear way of indicating the correct shapes. Do not award the mark if additional shapes are indicated, unless it is clear that the correct two shapes are the child's final choice.	G
25	Both inequalities completed correctly, using each of the given numbers once only, i.e. 28 > 18 48 > 38 38 > 18 48 > 28 38 > 28 48 > 18	1	Both inequalities must be correct for the award of the mark. **Do not** award the mark if any number is used more than once, or if numbers other than those given in the question are used.	E
26	29 (colouring pencils)	2	Award **2 marks** for the correct answer of 29 (colouring pencils). If the answer is incorrect or missing, award **1 mark** for evidence of a complete, correct method, e.g. • 50 – 12 – 9 = (incorrect **or** no answer) • 12 + 9 = 22 (*error*) 50 – 22 =	G
27	Number sentence completed as shown. 24 ÷ 2 = 12 pairs	1	All three numbers must be correct for the award of the mark.	E
28	Number sentence completed as shown. 14 + 7 = 10 + 11	1		E

Year 2 Summer Half Term 1: Reasoning: Mark Scheme

29	13 (bananas)	2	Award **2 marks** for the correct answer of 13. If the answer is incorrect or missing, award **1 mark** for evidence of a complete, correct method, e.g. 6 × 5 – 17 = (incorrect or no answer) 6 × 5 = 60 (error) 60 – 17 =	G
30	11 (p)	1		G
31	Any two numbers that are greater than 30 with a difference of 9, e.g. 60 – 51 45 – 36 44 – 35	1	Both numbers must be greater than 30 for the award of the mark. **Do not** award the mark if only one number is greater than 30 and has a difference of 9, e.g. 31 – 22. **Do not** award the mark if the numbers are presented in the wrong order, e.g. 36 – 45 = 9	G

Threshold scores
Working towards the expected standard (T): 20 or fewer
Working at the expected standard (E): 21–27
Working at greater depth (G): 28–35

Balance of difficulty of questions in the paper
8 marks at working towards (T)
19 marks at the expected standard (E)
8 marks at working at greater depth (G)

Year 2 Summer Half Term 2: Arithmetic Mark Scheme

Question	Requirement	Mark	Level of demand
1	5	1	T
2	6	1	T
3	90	1	T
4	55	1	T
5	12	1	E
6	24	1	T
7	65	1	T
8	12	1	E
9	40	1	T
10	75	1	E
11	21	1	G
12	45	1	E
13	100	1	E
14	1	1	E
15	62	1	E
16	101	1	E
17	12	1	E
18	29	1	G
19	5	1	E
20	36	1	E
21	7	1	G
22	53	1	G
23	21	1	G
24	83	1	G
25	10	1	G

Threshold scores
Working towards the expected standard: 14 or fewer
Working at the expected standard: 15–19
Working at greater depth: 20–25

Balance of difficulty of questions in the paper
7 marks at working towards (T)
11 marks at the expected standard (E)
7 marks at working at greater depth (G)

© HarperCollins*Publishers* Ltd 2019

Year 2 Summer Half Term 2: Reasoning Mark Scheme

Question	Requirement	Mark	Additional guidance	Level of demand
1	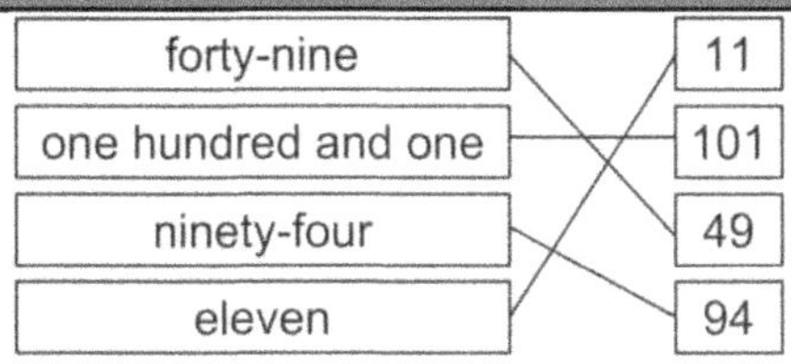	1	All **three** pairs of words and numbers must be correctly joined for award of the mark. Accept any other clear way of indicating the correct answer. **Do not** award the mark if other combinations are indicated unless it is clear the correct combination is the child's final choice.	T
2	Numbers put in the correct order as shown. 81 56 43 34 19 largest smallest	1	All numbers must be in the correct order for the award of the mark. Accept any other clear way of indicating the correct answer, e.g. matching each number to its correct position. Accept the numbers given in the reverse order, provided the labels have been swapped.	T
3	Numbers as below. 47 57 67	1	All numbers must be in the correct order for the award of the mark.	T
4	3 (cm)	1		T
5	Two sentences ticked as shown. A pentagon has 5 sides. ✓ A pentagon has curved sides. ☐ A regular pentagon has no lines of symmetry. ☐ A regular pentagon has sides of equal length. ✓	1	Both correct sentences must be indicated for the award of the mark. Accept any other clear way of indicating the correct sentences. **Do not** award the mark if other sentences are indicated unless it is clear that the correct sentences are the child's final choice.	T
6	The correct fraction ticked as shown. $\frac{1}{2}$ ☐ $\frac{1}{4}$ ✓ $\frac{1}{3}$ ☐ $\frac{3}{4}$ ☐	1	Accept any other clear way of indicating the correct answer. **Do not** award the mark if additional fractions are indicated, unless it is clear the correct fraction is the child's final choice.	E
7	(£)33	1	Accept any unambiguous indication of the correct amount, e.g. (£)33.00 (£)33.00p **Do not** award the mark for answers where the amount of pounds and/or pence are incorrect, e.g. (£)3300 or (£)3300p, unless the £ symbol has been crossed out in the answer box.	E
8a	90 (minutes)	1		E
8b	Hands drawn on clock to show quarter to 5. 	1	Allow slight inaccuracies in drawing the hands.	E
9	31 (°C)	1	Accept 32 °C. **Do not** accept 30 °C.	E

Year 2 Summer Half Term 2: Reasoning Mark Scheme

10	Number sentence completed as shown. 10 × 8 = 80 children or 8 × 10 = 80 children	1	All three numbers must be correct for the award of the mark.	E
11	Pairs matched as indicated: 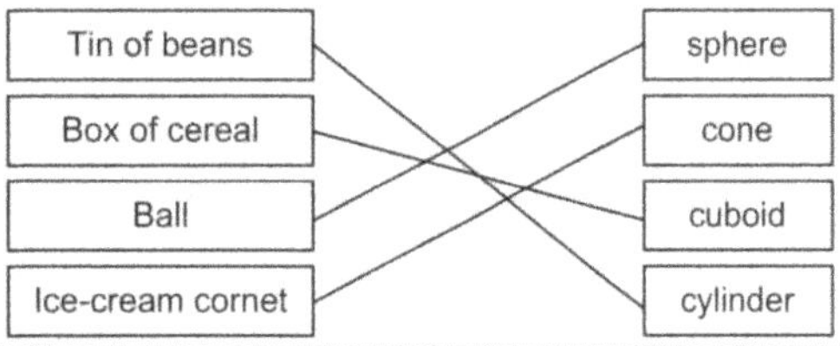	1	All three pairs must be correctly matched for the award of the mark. **Do not** award the mark if an object is matched to more than one shape.	T
12	25 (ml)	1		E
13	25 (shells)	1		E
14	72 (p)	1		E
15	43 (marbles)	1		E
16	Inequalities completed as shown. Pablo's mass < Percy's mass Patricia's mass > Pamela's mass	1		E
17	Two number sentences completed correctly, using four different number cards from those that are given, e.g. 10 + 70 = 80 20 + 60 = 80 30 + 50 = 80	2	Award **2 marks** for two number sentences completed correctly. Accept 70 + 10 = 80, 60 + 20 = 80 or 50 + 30 = 80. **Do not** accept 40 + 40 = 80. Accept any other clear way of indicating the correct answers, e.g. matching correct cards to answer boxes. Award **1 mark** if one number sentence is correct, but the child has used the same number cards for the other number sentence. Do not award any marks if the child uses numbers that are not given in the question.	E
18	Any combination of coins indicated that totals 55p, e.g. 50p + 5p (shown). 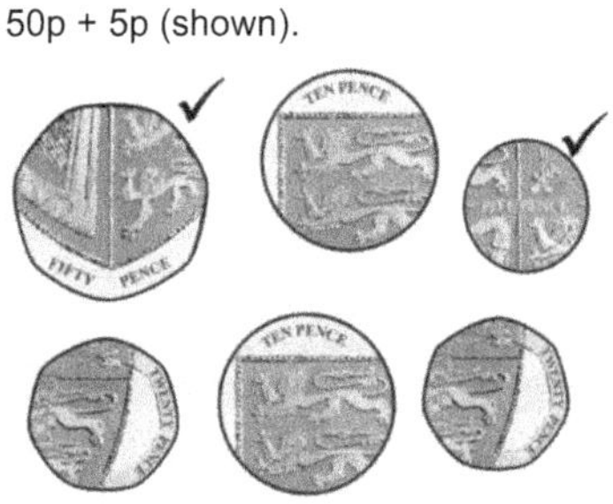 Or 20p, 20p, 10p, 5p	1	Award the mark for any other clear way of indicating a correct combination of coins, i.e. circling, underlining or drawing other coins.	E
19a	9 cm line drawn	1	Accept a line length of between $8\frac{1}{2}$ cm and $9\frac{1}{2}$ cm.	E
19b	9 (cm)	1		E

© HarperCollins*Publishers* Ltd 2019

Year 2 Summer Half Term 2: Reasoning Mark Scheme

20	Both number sentences completed as shown. One from: 6 × 3 = 18 3 × 6 = 18 and one from: 18 ÷ 3 = 6 18 ÷ 6 = 3	1	All numbers in both number sentences must be correct for the award of the mark.	E
21	Clock face ticked as shown. 	1	Accept any other clear way of indicating the correct clock face. **Do not** award the mark if other clock faces are indicated unless it is clear that the correct clock face is the child's final answer.	E
22	36 (kg)	1		E
23	Correct answer circled as shown. quarter turn, half turn, (three-quarter turn), full turn	1		E
24	20 (g)	1		G
25	Fraction completed as shown. $\frac{3}{4}$ of 44 = 33	1		G
26	Clockface showing 25 past 4. 	1	Accept slight inaccuracies in the drawing of the hands.	G
27	(£) 12	1		G
28	53	2 or 1	Award **2 marks** for the correct answer of 53. If the answer is incorrect or missing, award 1 mark for evidence of a complete, correct method, e.g. 8 × 10 – 27 = (incorrect or no answer) 8 × 10 = 90 (error) 90 – 27 =	G
29	Award the mark for any two numbers that are greater than 50 with a difference of 7, e.g. 77 – 70 60 – 53 59 – 52	1	Both numbers must be greater than 50 for the award of the mark. **Do not** award the mark if only one number is greater than 50 and the difference is 7, e.g. 57 – 50 **Do not** award the mark if the numbers are presented in the wrong order, e.g. 70 – 77 = 7	G
30	Number sentence completed as shown. 50 + 5 = 48 + 7	1		G

© HarperCollins*Publishers* Ltd 2019

31	Both inequalities completed correctly, using each of the given numbers once only, i.e. 21 < 22, 24 > 23. 21 < 23, 24 > 22. 21< 24, 23 > 22. 22 < 23, 24 > 21. 22 < 24, 23 > 21. 23 < 24, 22 > 21.	1	Both inequalities must be correct for the award of the mark. **Do not** award the mark if any number is used more than once, or if numbers other than those given in the question are used.	E

Threshold scores
Working towards the expected standard (T): 20 or fewer
Working at the expected standard (E): 21–27
Working at greater depth (G): 28–35

Balance of difficulty of questions in the paper
6 marks at working towards (T)
21 marks at the expected standard (E)
8 marks at working at greater depth (G)

© HarperCollins*Publishers* Ltd 2019

Content domain references

Autumn 1: Arithmetic	
Question	Content domain reference
1	1N1b/2N6
2	1N1b/2N6
3	1F1a/2F1b
4	1N2b
5	1C2a/2C2a
6	2N1
7	1N1b/2C6
8	2C2a
9	2C2a/2C3
10	2C2b/2N6
11	2C2b/1N1a
12	2C2a
13	1C4/2C1
14	2C1/2C3
15	2C1/2C3

Autumn 2: Arithmetic	
Question	Content domain reference
1	1N1b/2N6
2	2N6/2C2a
3	1C2a
4	2C6/1N1b
5	2N1/2C2a
6	2C3
7	2C2b
8	2C2b
9	2C8
10	2C6/1N1b
11	2C9a
12	2C4
13	2C3/1C2a
14	2C6
15	2C7/2C8

Autumn 1: Reasoning	
Question	Content domain reference
1	2N3/2N4
2	2N4
3	2N4
4	2N4
5	2N4
6	2N2a
7	2N1
8	2N4
9	2N2b/2N3
10	2N6/2C2a
11	2N1/2N4
12	2C4
13	2N6
14	2N2b
15	2N4
16	2C4
17	2C1/2C3
18	2C9a

Autumn 2: Reasoning	
Question	Content domain reference
1	1C1/1C4
2	2C1/2C4
3	2N6/2M3b
4	1C8/1N1b
5	2M9/2M3a
6	2C8
7	2C6
8	1N1b/1C8
9	1C4
10	2C4
11	2C7
12	1C1/1C4
13	2N2b/2C6
14	2C2a/2C2b
15	2N2b/2C4
16	2M3a/2M9
17	2C4
18	2C8/2M3b
19	2M9
20	2C9b
21	2C8/2M9
22	2C8/2C9b
23	2C3/2C2b

© HarperCollinsPublishers Ltd 2019

Content domain references

Spring 1: Arithmetic	
Question	Content domain reference
1	1N1b/2C1
2	2N1/2C2a
3	2C6/1N1b
4	2C6/1N1b
5	2C2a/2C2b
6	2C6/1N1b
7	1C1/1C2a
8	2C6
9	2C2a/2C2b
10	1C4/2C1
11	2C6/1N1b
12	2C2a/2C2b
13	1C4
14	2C6/2C9b
15	2C2a/2C2b
16	2C3/2C2b
17	2N1/1N1a
18	2C9a
19	2C3/2C2b
20	2C7

Spring 1: Reasoning	
Question	Content domain reference
1	1C8/2C7
2	2N6/1N4
3	1G1a
4	1C8
5	2N6/2C2a
6	2C4
7a	2M3b/2M9
7b	2M3b/2M9
8	2C9b
9	1N1b/2C4
10	2C6/2C7
11	2C2a/2C4
12	2M3a/2M9
13a	2S1/2S2b
13b	2S1/2S2a
14	1G1a
15	2M3a/2M9
16	2N2b/2C2a
17	2C4
18	2M3a/2M9
19	2G2b/2G3
20	2C6/2C7
21	2C4
22	2C3

Spring 2: Arithmetic	
Question	Content domain reference
1	1N1b
2	1N1b/2N6
3	2C6/1N1b
4	1C4/2C1
5	1C1/1C2a
6	1F1a/2F1a
7	2C2a/2C2b
8	2C2a/2C2b
9	2C6/1N1b
10	2C6
11	2F1a/1F1a
12	2C2a/2C2b
13	2C6
14	2C6
15	2C3/2N1
16	2C9a
17	2C6
18	2C3/2C2b
19	2C7
20	2F1a/2F1b

Spring 2: Reasoning	
Question	Content domain reference
1	2C6
2	1M1
3	2F1a
4	1G1b/2G2b
5	2C4
6	1F1b/2F1a
7	2C4
8a	2S2b
8b	2S2a
9	2F1a
10	2C4
11	2M3a/2M9
12	2C4
13	1C8/2C6
14	2G2a
15	2M2
16	2C6/2C7
17	2F1a
18	2C8/2M3b
19	2M2/1G1a
20	2F1a/2F2
21	2C9b
22	2M9/2M3a
23	2C4
24	2M1
25	2C4
26	2C3/2C2b
27	2C3

© HarperCollins*Publishers* Ltd 2019

Content domain references

Summer 1: Reasoning	
Question	**Content domain reference**
1	2N2a
2	2N1/2N6
3	1C1
4	2C4
5	2C6/2C8
6	2C4
7	2M4a
8	2C4
9a	2G3
9b	2P1
10	2F2/1F1a
11	1C8/2C8
12	2M3b/1N1b
13	2M4c
14	1P2/2P2
15a	2S1/2S2a
15b	2S1/2S2b
16	2F1a
17	2M4a
18	2C4
19	2F1b/2F2
20	2C2a/2C2b
21	1M1/2M2
22	2M9
23	2C7
24	2G2b
25	2N2b
26	2C4
27	2C6/2C7
28	2C4
29	2C8/2C4
30	2F1a/2F1b
31	2C3/2C4

Summer 1: Arithmetic	
Question	**Content domain reference**
1	1C2a
2	2N6
3	2C2a/2C2b
4	2C2a/2C2b
5	2C6/1N1b
6	2C1
7	1C4
8	2C1/2C3
9	2F1a
10	2C1/2C3
11	2C2a/2C2b
12	2C6
13	2C1
14	2C2a/2C2b
15	2C6
16	2C2a/2C2b
17	2F1a
18	2C6
19	2C2a/2C2b
20	2C7
21	2C3
22	2C6
23	2C3
24	2F1a
25	2C3

Summer 2: Arithmetic	
Question	**Content domain reference**
1	1C2a
2	1C4/2C1
3	2N6/1N1b
4	2C2a/2C2b
5	2C1
6	2C6/1N1b
7	2C2a/2C2b
8	2F1a
9	2C1/2C3
10	2C2a/2C2b
11	2C7/2N1
12	2C2a/2C2b
13	2C1
14	2C6
15	2C2a/2C2b
16	2C2a/2C2b
17	2C6
18	2C3
19	2F1a
20	2C2a/2C2b
21	2C6
22	2C3
23	2F1a
24	2C3
25	2C6

Summer 2: Reasoning	
Question	**Content domain reference**
1	2N2a/2N3
2	2N2b/2N3
3	1N1b/2N2a
4	1M2/2M2
5	2G2a
6	2N4/2F1a
7	2C4/2M9
8a	2M4b/2M4c
8b	2M4a
9	2M2
10	2C7/2C6
11	1G1b/2G1b
12	2C4/2M1
13	2N1/2S2b
14	2M9
15	2C4
16	2M1
17	2C1/2C2b
18	2M9/2M3a
19a	1M2/2F1a
19b	2M2
20	2C9b
21	2M4a
22	2C2a/2M2
23	2P2
24	2C6/2C8
25	2F1a/2F1b
26	2M4a/2M4c
27	2F1a/2M9
28	2C8/2C4
29	2C3/2C4
30	2C4/2C9a
31	2N2b

© HarperCollins*Publishers* Ltd 2019

Name ______________________ Class ____________

Year 2/P3 Maths Progress Tests for White Rose Record Sheet

Tests	Mark	Total marks	Key skills to target
Autumn 1: Arithmetic			
Autumn 1: Reasoning			
Autumn 2: Arithmetic			
Autumn 2: Reasoning			
Spring 1: Arithmetic			
Spring 1: Reasoning			
Spring 2: Arithmetic			
Spring 2: Reasoning			
Summer 1: Arithmetic			
Summer 1: Reasoning			
Summer 2: Arithmetic			
Summer 2: Reasoning			

© HarperCollins*Publishers* Ltd 2019